Introduction

Blake Hudson

Welcome to the wonderful world of TROPICAL FOOD PLANTS! It is my passion to explore the incredible diversity of edible treasures that can be grown in the tropics. For nearly two decades, I have been cultivating and researching these plants, and now I am thrilled to share my wealth of knowledge with you.

As a permaculture designer, I know firsthand the importance of having a comprehensive, user-friendly guide to help plan and execute a thriving food garden. After scouring through countless books on tropical fruits and plants, I decided to create a system that would provide clear and concise information in an easy-to-read format.

My passion for these plants extends beyond their individual qualities, as I am dedicated to creating a harmonious ecosystem that is both productive and sustainable. That's why I have included information on food forests and banana circles, which are crucial elements that help tie all the plants listed together in a way that is thriving and interconnected.

I hope this guide inspires you to embrace the tropical bounty and embark on a journey of discovery and fulfillment. May your gardens flourish and your plates overflow with the vibrant and delicious fruits of your labour. Happy productive gardening!

How to use this guide

Image of the fruit or most identifying feature

Common name

Botanical name

Type of plant

Size width, height & spacing between the same species

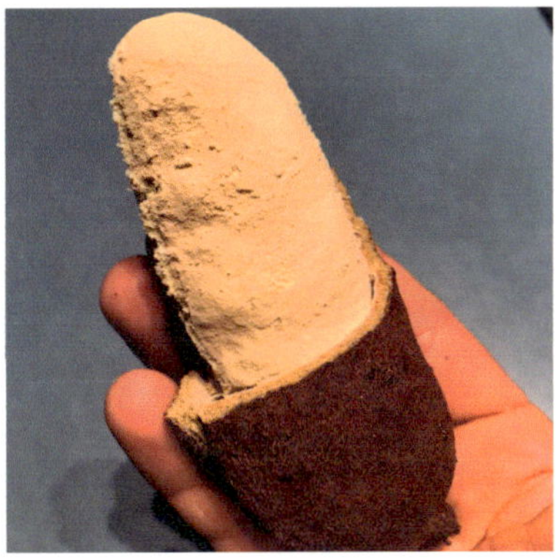

WEST INDIAN LOCUST, *Hymenaea courbaril*

- Large tree
- Produces an abundant crop after 8 years
- Tastes like carob
- Used in treats and smoothies
- Fruit eaten fresh
- High in protein
- Leaves are cooked in soup

SHAPE			
TREE	PALM	SHRUB	CLUMP
VINE	HERB	GRASS	GROUND COVER

GROWTH (METRES)					
H	25	W	15	SPACE	15

PRODUCTIVE SEASON											
J	F	M	A	M	J	J	A	S	O	N	D

LIGHT REQUIREMENTS		
SHADE	PART SHADE	FULL SUN

Most useful tips and info you will need

Which months it's most productive in the southern hemisphere

Light requirements for ideal growth and fruiting ability. This also can indicate its layer or strata in the forest. For example less light requirements means it can grow the lower layers

ACAI, *Euterpe oleracea*

- Clumping palm.
- Produces small, dark purple fruits in large clusters.
- Fruits are rich in antioxidants, healthy fats, and fibre.
- Take 4-8 years to fruit.
- Propagated by seeds, prefers fertile, well-draining soil.

SHAPE			
TREE	**PALM**	SHRUB	CLUMP
VINE	HERB	GRASS	GROUND COVER

GROWTH (METRES)					
H	10	W	5	SPACE	7

PRODUCTIVE SEASON											
J	**F**	M	A	M	J	J	A	S	O	N	**D**

LIGHT REQUIREMENTS		
SHADE	**PART SHADE**	**FULL SUN**

ACKEE, *Blighia sapida*

- Eaten only after the fruit has opened naturally on the tree, unripe fruit is poisonous.
- Relative to lychee.
- Fruit has a cheesy nut butter flavour
- Can be kept small in urban areas
- 3 plants required for pollination

SHAPE			
TREE	PALM	SHRUB	CLUMP
VINE	HERB	GRASS	GROUND COVER

GROWTH (METRES)					
H	10	W	5	SPACE	7

PRODUCTIVE SEASON											
J	**F**	**M**	**A**	**M**	J	J	A	S	O	N	D

LIGHT REQUIREMENTS		
SHADE	**PART SHADE**	**FULL SUN**

ABIU, *Pouteria caimito*

- The abiu is a smooth, brilliant yellow fruit with sweet translucent flesh that is very delicious.
- In Brazil the pulp from the plant is used medicinally, to relieve coughs and bronchitis.
- Medium sized tree with light foliage.

SHAPE			
TREE	PALM	SHRUB	CLUMP
VINE	HERB	GRASS	GROUND COVER

GROWTH (METRES)					
H	4	W	3	SPACE	3

PRODUCTIVE SEASON											
J	**F**	**M**	A	M	**J**	**J**	A	S	O	N	D

LIGHT REQUIREMENTS		
SHADE	**PART SHADE**	**FULL SUN**

ACEROLA, *Malpighia emarginata*

- This small bright red cherry-like fruit has varying tastes during different stages in maturity.
- When fully ripe the fruit is juicy and aromatic with its acid content giving it a sweet acid apple-like flavour.
- Multiple yearly crops.
- Wide tree with attractive cascading branches.

SHAPE			
TREE	PALM	**SHRUB**	CLUMP
VINE	HERB	GRASS	GROUND COVER

GROWTH (METRES)					
H	4	W	4	SPACE	4

PRODUCTIVE SEASON											
J	**F**	**M**	**A**	**M**	J	J	A	S	**O**	**N**	**D**

LIGHT REQUIREMENTS		
SHADE	**PART SHADE**	FULL SUN

ACHACHA, Garcinia humilis

- Achachas are tangy and refreshing with a delicate subtle sweetness, a fine balance between sweetness and acidity
- Should be picked mature as it does not ripen further on storage
- Evergreen
- Suitable trees for suburban areas and can be kept to manageable size without much pruning
- Time to fruit 5 years

SHAPE			
TREE	PALM	SHRUB	CLUMP
VINE	HERB	GRASS	GROUND COVER

GROWTH (METRES)					
H	4	W	3	SPACE	3

J	**F**	**M**	**A**	M	J	J	A	S	O	N	D

LIGHT REQUIREMENTS		
SHADE	**PART SHADE**	FULL SUN

ATHERTON RASPBERRY, *Rubus probus*

- Rambling vine with spikes.
- Best kept as pruned on a trellis otherwise it can take over the garden bed.
- Prune after fruiting.
- Best flavoured berries are picked when the fruit basically falls into the hand.
- Propagated from cutting or root division.
- Fruit within 1-2 years.

SHAPE			
TREE	PALM	**SHRUB**	CLUMP
VINE	HERB	GRASS	GROUND COVER

GROWTH (METRES)					
H	1	W	3	SPACE	1

PRODUCTIVE SEASON											
J	F	**M**	A	**M**	J	**J**	A	**S**	O	**N**	D

LIGHT REQUIREMENTS		
SHADE	PART SHADE	FULL SUN

AVOCADO, *Persea americana*

- In wet tropics with clay soil plant into a large mound about 3m wide x .5m deep to avoid root disease.
- Fruit from 2-3 years when grafted
- Varieties with scattered fruiting times:
  - Sheppard - Feb to April, Hass - Sept to Jan, Wurts - June to Oct, Rincon - July to Sept, Bacon - June to July

SHAPE			
TREE	PALM	SHRUB	CLUMP
VINE	HERB	GRASS	GROUND COVER

GROWTH (METRES)					
H	4	W	3	SPACE	3

PRODUCTIVE SEASON											
J	**F**	**M**	**A**	M	**J**	**J**	**A**	**S**	**O**	**N**	**D**

LIGHT REQUIREMENTS		
SHADE	PART SHADE	**FULL SUN**

BANANA, *Musa sp.*

- Many types eaten freash or cooked,
- Each type is quite different and texture, colour and flavour.
- Peak season is the wet season, although you may find a bunch to harvest most times of the year.
- Multi-purpose plant, food, fibre and medicine while being great soil improver.
- Fruit within 1 year.

SHAPE			
TREE	PALM	SHRUB	**CLUMP**
VINE	**HERB**	GRASS	GROUND COVER

GROWTH (METRES)					
H	3-5	W	3-5	SPACE	3

PRODUCTIVE SEASON											
J	**F**	**M**	A	M	J	J	A	**S**	**O**	**N**	**D**

LIGHT REQUIREMENTS		
SHADE	**PART SHADE**	**FULL SUN**

BILIMBI, *Averrhoa bilimbi*

- Closely related to the carambola with crispy fruit that is very acidic.
- Generally not eaten fresh but used as a vegetable in cooking and making relish.
- Grown from seedling.
- Can be pruned to shape.
- Fruit in 5 years.
- Requires frequent watering.
- Evergreen.

SHAPE			
TREE	PALM	SHRUB	CLUMP
VINE	HERB	GRASS	GROUND COVER

GROWTH (METRES)					
H	4	W	3	SPACE	3

PRODUCTIVE SEASON											
J	**F**	**M**	**A**	**M**	**J**	J	A	S	O	**N**	**D**

LIGHT REQUIREMENTS		
SHADE	PART SHADE	**FULL SUN**

BLACK MULBERRY, *Morus nigra*

- Sweet fruit can reach up to 5cm in length.
- Delicious eaten fresh but can also be dried and eaten as a sweet.
- Medium sized, spreading tree with a weeping habit, excellent shade tree.
- Can be pruned or branches weaved to provide a shelter.
- Deciduous over winter.
- Propagated from cutting.

SHAPE			
TREE	PALM	SHRUB	CLUMP
VINE	HERB	GRASS	GROUND COVER

GROWTH (METRES)					
H	4	W	4	SPACE	4

PRODUCTIVE SEASON											
J	F	M	A	M	J	J	A	**S**	**O**	**N**	D

LIGHT REQUIREMENTS		
SHADE	**PART SHADE**	**FULL SUN**

BLACK SAPOTE, *Diospyros nigra*

- Large hard green fruit that turn very soft and black when ripe; tasting like creamy chocolate mousse.
- Wind tolerant
- Frost sensitive
- Pick fruit when the skirt on top begins to curl
- Will grow into a large tree if allowed

SHAPE			
TREE	PALM	SHRUB	CLUMP
VINE	HERB	GRASS	GROUND COVER

GROWTH (METRES)					
H	10	W	6	SPACE	6

PRODUCTIVE SEASON											
J	F	M	A	**M**	**J**	**J**	A	S	O	N	D

LIGHT REQUIREMENTS		
SHADE	PART SHADE	**FULL SUN**

BLUEBERRY, *Biloxi vaccinium*

- Small delicate plant that doesn't tolerate the heat.
- Best suited to sub-tropical to temperate climate.
- Prefers slightly acidic soils; 5.5-6.5 PH; therefore do not add lime/dolomite.
- Shallow rooted, so mulch well.
- Does not tolerate soggy soil, commercial grown in pots.

SHAPE			
TREE	PALM	**SHRUB**	CLUMP
VINE	HERB	GRASS	GROUND COVER

GROWTH (METRES)					
H	1.2	W	1	SPACE	1

PRODUCTIVE SEASON											
J	**F**	**M**	**A**	M	J	J	**A**	**S**	**O**	**N**	**D**

LIGHT REQUIREMENTS		
SHADE	PART SHADE	FULL SUN

BREADFRUIT, *Artocarpus altilis*

- Tropical potato that grows on a tree.
- Eaten unripe and cooked as a potato.
- Allowed to ripen it's eaten as a dessert.
- Can grow very tall.
- Wind sensitive.
- Propagated by a slight cut of a root and allowing the cut to grow a new plant.
- Will fruit withing 4-5 years.

SHAPE			
TREE	PALM	SHRUB	CLUMP
VINE	HERB	GRASS	GROUND COVER

GROWTH (METRES)					
H	10	W	8	SPACE	8

PRODUCTIVE SEASON											
J	**F**	M	A	M	J	J	A	S	O	N	D

LIGHT REQUIREMENTS		
SHADE	PART SHADE	**FULL SUN**

CACAO, *Theobroma cacao*

- Seed is used to create chocolate after fermentation and processing.
- Fruit around the seed has a sweet tangy flavour.
- Medium sized evergreen tree with large green leaves.
- Propagate by seed.
- Fruit within 3-4 years.

SHAPE			
TREE	PALM	SHRUB	CLUMP
VINE	HERB	GRASS	GROUND COVER

GROWTH (METRES)					
H	4	W	3	SPACE	4

PRODUCTIVE SEASON											
J	F	M	A	M	J	J	A	S	**O**	**N**	**D**

LIGHT REQUIREMENTS		
SHADE	PART SHADE	FULL SUN

CANISTEL, *Pouteria campechiana*

- The yellow flesh has a sweet earthy flavour and a meaty texture similar to a boiled egg
- Aztec dish which is like guacamole but with Canistel.
- Versitile fruit can be eaten fresh or used in desserts or cakes.
- Rich in vitamins A, C, and niacin, it supports eye health and immunity.
- Will fruit within 4 years.

SHAPE			
TREE	PALM	SHRUB	CLUMP
VINE	HERB	GRASS	GROUND COVER

GROWTH (METRES)					
H	5	W	5	SPACE	5

PRODUCTIVE SEASON											
J	F	M	A	M	J	J	A	S	**O**	**N**	**D**

LIGHT REQUIREMENTS		
SHADE	PART SHADE	**FULL SUN**

CAPE GOOSEBERRY, *Physalis peruviana*

- Delicate shrub that grows a sweet citrusy fruit.
- Low growing and sprawling plant.
- Fruits most of the year.
- Can grow in very acid and very alkaline soils.
- Prefers well-drained soil and can grow in nutritionally poor soil.
- Fruit within first year.

SHAPE			
TREE	PALM	**SHRUB**	CLUMP
VINE	HERB	GRASS	GROUND COVER

GROWTH (METRES)					
H	1.5	W	2	SPACE	.5

PRODUCTIVE SEASON											
J	**F**	M	**A**	M	**J**	J	**A**	S	**O**	N	**D**

LIGHT REQUIREMENTS		
SHADE	PART SHADE	FULL SUN

CAPE PLUM, *Flacourtia sp.*

- Native to northern Australia.
- Large shrub.
- Masses of black berries that are sweet and can be used to make jam.
- Male and female plants needed so its best to have at lease two plants, to ensure you have both genders to produce fruit.
- Fruit within 3-4 years.

SHAPE			
TREE	PALM	**SHRUB**	CLUMP
VINE	HERB	GRASS	GROUND COVER

GROWTH (METRES)					
H	3	W	2	SPACE	3

PRODUCTIVE SEASON											
J	**F**	**M**	**A**	**M**	**J**	J	A	S	O	N	D

LIGHT REQUIREMENTS		
SHADE	**PART SHADE**	FULL SUN

CARAMBOLA, *Averrhoa carambola*

- Sweet juicy start shaped fruit.
- Evergreen foliage
- Pink flowers
- Will grow tall and slender if not pruned
- Leaves and fruit are good animal fodder
- Will fruit within 2-4 years

SHAPE			
TREE	PALM	SHRUB	CLUMP
VINE	HERB	GRASS	GROUND COVER

GROWTH (METRES)					
H	5	W	5	SPACE	5

PRODUCTIVE SEASON											
J	**F**	**M**	A	M	J	J	A	**S**	**O**	**N**	**D**

LIGHT REQUIREMENTS		
SHADE	PART SHADE	**FULL SUN**

CASHEW, *Anacardium occidentale*

- Nuts require roasting, but be warned shell contains caustic oil which can burn the skin so be careful.
- The fruit is external to the nut and is highly aromatic and sweet.
- Will not tolerate poorly drained soils.
- Attractive tree with beautiful flowers.
- Protect from strong winds.

SHAPE			
TREE	PALM	SHRUB	CLUMP
VINE	HERB	GRASS	GROUND COVER

GROWTH (METRES)					
H	6	W	6	SPACE	6

PRODUCTIVE SEASON											
J	**F**	**M**	A	M	J	J	A	S	**O**	**N**	**D**

LIGHT REQUIREMENTS		
SHADE	PART SHADE	**FULL SUN**

CEDAR BAY CHERRY, *Eugenia reinwardtiana*

- Northern Australian native
- Small red edible fruiting shrub.
- Slow growing.
- Produces a tangy and sweet tasty red fruit.
- Susceptible to myrtle rust (*Pussinia psidii*).
- Fertilize with compost to speed up growth.
- Fruit within 2-3 years.

SHAPE			
TREE	PALM	**SHRUB**	CLUMP
VINE	HERB	GRASS	GROUND COVER

GROWTH (METRES)					
H	1	W	2	SPACE	.5

PRODUCTIVE SEASON											
J	**F**	**M**	A	M	J	J	A	S	O	N	D

LIGHT REQUIREMENTS		
SHADE	PART SHADE	FULL SUN

CITRUS sp.

- All citrus grow very well in the tropics.
- Prune yearly to maintain shape.
- Best grown from a grafted tree but can do well from seed and layering.
- Many varieties can be grafted to one tree.
- Reliable market value and shelf life.
- Can be grown from seed or graft.
- Fruit within 2-3 years.

SHAPE			
TREE	PALM	**SHRUB**	CLUMP
VINE	HERB	GRASS	GROUND COVER

GROWTH (METRES)					
H	3	W	3	SPACE	3

PRODUCTIVE SEASON											
J	F	**M**	**A**	**M**	**J**	**J**	A	S	O	N	D

LIGHT REQUIREMENTS		
SHADE	**PART SHADE**	FULL SUN

CLOVE, Syzygium aromaticum

- Native to Indonesia and Philippines
- Not water demanding
- Evergreen
- When grown from seed takes 5-10 years to flower
- Cuttings can be taken for propagation
- The aromatic flower bud is dried and use as spice

SHAPE			
TREE	PALM	**SHRUB**	CLUMP
VINE	HERB	GRASS	GROUND COVER

GROWTH (METRES)					
H	1.5	W	1.2	SPACE	1

PRODUCTIVE SEASON											
J	**F**	**M**	A	M	J	J	A	S	**O**	**N**	**D**

LIGHT REQUIREMENTS		
SHADE	**PART SHADE**	**FULL SUN**

COCONUT, *Cocos nucifera*

- Many varieties however Malay Gold bears golden fruit, water and flesh are beautifully sweet.
- Very nutritious in all stages of maturity.
- Free draining soil and are best planted deep with a deep organic mulch or coconut husk surrounding them.
- Sea salt / water helps with fruiting.
- Fruit within 3-4 years.

SHAPE			
TREE	**PALM**	SHRUB	CLUMP
VINE	HERB	GRASS	GROUND COVER

GROWTH (METRES)					
H	5	W	4	SPACE	4

PRODUCTIVE SEASON											
J	**F**	**M**	A	M	J	J	A	S	O	N	D

LIGHT REQUIREMENTS		
SHADE	**PART SHADE**	**FULL SUN**

COFFEE, Coffea arabica

- 9 plants are required to produce enough beans for 1 espresso shot per day.
- Small evergreen dense shrub.
- Prefers rich, well-drained soil and partial shade with consistent rainfall.
- Can become weedy.
- Propagated by seed or cuttings; begins fruiting in 3-4 years.

SHAPE			
TREE	PALM	**SHRUB**	CLUMP
VINE	HERB	GRASS	GROUND COVER

GROWTH (METRES)					
H	3	W	2	SPACE	1

PRODUCTIVE SEASON											
J	F	M	A	M	**J**	**J**	**A**	**S**	O	N	D

LIGHT REQUIREMENTS		
SHADE	**PART SHADE**	FULL SUN

CUSTARD APPLE, Annona atemoya

- Meduim size wide and sprawling canopy.
- Grows abundant delicious very sweet fruit.
- Grow best in fertile soil but will grow well in poorer soils with proper attention to nutrition.
- Hybrid to the cherimoya which grows in cooler climates.
- Fruit within 4-6 years from seedling.

SHAPE			
TREE	PALM	SHRUB	CLUMP
VINE	HERB	GRASS	GROUND COVER

GROWTH (METRES)					
H	4	W	4	SPACE	4

PRODUCTIVE SEASON											
J	F	M	A	M	**J**	**J**	**A**	**S**	**O**	**N**	D

LIGHT REQUIREMENTS		
SHADE	PART SHADE	**FULL SUN**

DAVIDSON PLUM, *Davidsonia pruriens*

- Native to north QLD Australia.
- Evergreen rainforest tree with large, dark green, leathery, divided leaves.
- Grows best in full sun to part shade.
- Plum like berries are very sour which make great jam.
- Sliced fruit in a jug of water make a refreshing drink.
- Fruit within 4-5 years from seed.

SHAPE			
TREE	PALM	SHRUB	CLUMP
VINE	HERB	GRASS	GROUND COVER

GROWTH (METRES)					
H	5	W	2	SPACE	2

PRODUCTIVE SEASON											
J	**F**	**M**	**A**	**M**	J	J	A	S	O	N	D

LIGHT REQUIREMENTS		
SHADE	**PART SHADE**	**FULL SUN**

DRAGON FRUIT RED, *Hylocereus costaricensis*

- Best grown on a Timber log with a X at the top that gives support and allows it to hang down to fruit.
- Many varieties, including red with red flesh, red with white flesh, yellow with white flesh and even purple!
- Fruit within 1-2 years from cutting.

SHAPE			
TREE	PALM	SHRUB	CLUMP
VINE	HERB	GRASS	GROUND COVER

GROWTH (METRES)					
H	-	W	2	SPACE	3

PRODUCTIVE SEASON											
J	**F**	**M**	A	M	J	J	A	S	O	N	D

LIGHT REQUIREMENTS		
SHADE	**PART SHADE**	FULL SUN

DURIAN, *Durio Zebethinus*

- Durian trees have one or two flowering and fruiting periods per year.
- The durian fruit can hang from any branch and matures roughly three months after pollination.
- For the first 2 years they require dense shade.
- Fruit within 4-5 years.

SHAPE			
TREE	PALM	SHRUB	CLUMP
VINE	HERB	GRASS	GROUND COVER

GROWTH (METRES)					
H	5	W	4	SPACE	4

PRODUCTIVE SEASON											
J	**F**	**M**	A	M	J	J	A	S	O	N	D

LIGHT REQUIREMENTS		
SHADE	PART SHADE	**FULL SUN**

ELDER BERRY, *Sambucus nigra*

- Medicinal plant which the flowers and fruits can be processed into drinks, jams and syrups as strong antiviral.
- The stem is hollow useful for flutes.
- Has lateral roots that spread easily.
- Kept pruned as a ornamental hedge.
- Flower and fruit attract birds and insects.
- Mineral accumulator, great in compost.

SHAPE			
TREE	PALM	**SHRUB**	CLUMP
VINE	HERB	GRASS	GROUND COVER

GROWTH (METRES)					
H	2.5	W	2	SPACE	1

PRODUCTIVE SEASON											
J	**F**	**M**	**A**	**M**	**J**	**J**	**A**	**S**	**O**	**N**	**D**

LIGHT REQUIREMENTS		
SHADE	**PART SHADE**	FULL SUN

FEIJOA, *Feijoa sellowiana*

- Requires two plants for pollination.
- Related to guava but with superior qualities in flavour and usage.
- Prefers sub-tropical climates.
- Can be pruned to keep compact.
- Suits most soil types.
- Fruit within 4-5 years from seedling.

SHAPE			
TREE	PALM	**SHRUB**	CLUMP
VINE	HERB	GRASS	GROUND COVER

GROWTH (METRES)					
H	3	W	2.5	SPACE	2

PRODUCTIVE SEASON											
J	F	**M**	**A**	**M**	**J**	J	A	S	O	N	D

LIGHT REQUIREMENTS		
SHADE	**PART SHADE**	FULL SUN

FINGER LIME, *Citrus australasica*

- High in vitamin c.
- Native to north QLD Australia.
- Spikey small shrub.
- The elongated cylindrical fruits may be green, yellow, black, brown or purple producing flesh that is green, yellow or pink.
- Fruit within 5+ years.

SHAPE			
TREE	PALM	**SHRUB**	CLUMP
VINE	HERB	GRASS	GROUND COVER

GROWTH (METRES)					
H	1.5	W	1	SPACE	1

PRODUCTIVE SEASON											
J	**F**	**M**	**A**	**M**	**J**	**J**	A	S	O	N	**D**

LIGHT REQUIREMENTS		
SHADE	**PART SHADE**	FULL SUN

GRAPE, *Vitis vinifera*

- Deciduous vine.
- Prune stems 3 nodes after the bunch in summer to help fruiting along with any unwanted leafy growth.
- Grown from cutting.
- Many varieties include Crimson, Pink, Black, Red.
- Heavy prune in winter.
- Best suited to dry tropics.

SHAPE			
TREE	PALM	SHRUB	CLUMP
VINE	HERB	GRASS	GROUND COVER

GROWTH (METRES)					
H	3	W	1	SPACE	1

PRODUCTIVE SEASON											
J	F	M	A	M	J	**J**	**A**	**S**	**O**	**N**	D

LIGHT REQUIREMENTS		
SHADE	**PART SHADE**	**FULL SUN**

GRUMICHAMA, *eugenia brasiliensis*

- 2 varieties; Purple or Yellow.
- Beautiful small tree has very attractive dark glossy-green waxy leaves and a peeling multi-coloured bark.
- Fruit has a sweet juicy cherry flavour.
- Can be grown as a hedge.
- Keep pruned as a bush.
- Fruit within 2-3 years from seed.

SHAPE			
TREE	PALM	**SHRUB**	CLUMP
VINE	HERB	GRASS	GROUND COVER

GROWTH (METRES)					
H	5-10	W	4	SPACE	2

PRODUCTIVE SEASON											
J	F	**M**	**A**	**M**	J	J	A	S	O	N	D

LIGHT REQUIREMENTS		
SHADE	**PART SHADE**	FULL SUN

GUAVA, *Psidium guajava*

- Many varieties; including the strawberry guava, white & yellow.
- Can be pruned to keep compact with multiple plantings as edible hedge.
- Grows well over a wide climatic range.
- Can become weedy as birds easily distribute seeds.
- Fruit within 2 years.

SHAPE			
TREE	PALM	**SHRUB**	CLUMP
VINE	HERB	GRASS	GROUND COVER

GROWTH (METRES)					
H	3	W	2.5	SPACE	2

PRODUCTIVE SEASON											
J	**F**	**M**	**A**	**M**	**J**	J	A	S	O	N	D

LIGHT REQUIREMENTS		
SHADE	**PART SHADE**	FULL SUN

HERBERTON PLUM, *Prunus salicinia*

- Japanese plum that has now been cultivated in Herberton for its hardiness in warmer climates
- Keep thinned to help grow bigger fruits
- Can handle frosts
- Fruits after 2-3 years

SHAPE			
TREE	PALM	SHRUB	CLUMP
VINE	HERB	GRASS	GROUND COVER

GROWTH (METRES)					
H	2	W	1.5	SPACE	2

PRODUCTIVE SEASON											
J	**F**	**M**	A	M	J	J	A	S	O	N	D

LIGHT REQUIREMENTS		
SHADE	**PART SHADE**	**FULL SUN**

HILL GOOSEBERRY, *Rhodomyrtus tomentosa*

- Ornamental shrub that grows a sweet fruit with similar taste of blueberry and raspberry.
- Frost hardy.
- Fire hardy.
- Weed potential.
- Great as a hedge.
- Fruit within 2-3 years from cutting.

SHAPE

TREE	PALM	**SHRUB**	CLUMP
VINE	HERB	GRASS	GROUND COVER

GROWTH (METRES)

H	1.5	W	2	SPACE	1

PRODUCTIVE SEASON

J	F	**M**	**A**	**M**	**J**	J	A	S	O	N	D

LIGHT REQUIREMENTS

SHADE	**PART SHADE**	FULL SUN

JABOTICABA YELLOW, *Myrciaria glazioviana*

- Yellow fruit sweet but slight pineapple/ mango flavour.
- Dense foliage.
- A single tree will produce fruit although cross pollination has shown to increase productivity.
- Fruit within 3-4 years.

SHAPE

TREE	PALM	**SHRUB**	CLUMP
VINE	HERB	GRASS	GROUND COVER

GROWTH (METRES)

H	3-5	W	3	SPACE	3

PRODUCTIVE SEASON

J	F	M	A	M	J	J	A	S	**O**	**N**	**D**

LIGHT REQUIREMENTS

SHADE	**PART SHADE**	**FULL SUN**

JABOTICABA, *Myrciara cauliflora*

- Fruit is similar to a grape with a sweet and aromatic flavour.
- Bears heavy crop of fruit 3 times a year
- Should NEVER be pruned as this will stop fruiting.
- Dense foliage great for screening.
- Makes great jam and wine.
- Multiple crops per year.
- Fruit within 6-9 years.

SHAPE

TREE	PALM	**SHRUB**	CLUMP
VINE	HERB	GRASS	GROUND COVER

GROWTH (METRES)

H	10	W	3	SPACE	3

PRODUCTIVE SEASON

J	F	**M**	A	**M**	J	**J**	A	S	O	N	**D**

LIGHT REQUIREMENTS

SHADE	**PART SHADE**	FULL SUN

JAKFRUIT, *Artocarpus heterophyllus*

- Super sweet fruit when ripe, unripe its cooked as a vegetable in curry.
- Can have latex skin, use oil on hands and knife before handling to avoid this.
- Seed can be used like potato, after peeling and boiling.
- Will grow tall quickly, so keep pruned.
- Fruit within 3 years.

SHAPE

TREE	PALM	SHRUB	CLUMP
VINE	HERB	GRASS	GROUND COVER

GROWTH (METRES)

H	10	W	3	SPACE	3

PRODUCTIVE SEASON

J	**F**	**M**	A	M	J	J	A	S	O	N	D

LIGHT REQUIREMENTS

SHADE	PART SHADE	**FULL SUN**

KAFFIR LIME, *Citrus Hystrix*

- Citrus tree mainly used for its aromatic leaves and fruit zest
- Flavouring common in Thai dishes.
- Traditional medicine in Asian countries.
- Leaves can be used all year round and fruiting is during winter.

SHAPE

TREE	PALM	**SHRUB**	CLUMP
VINE	HERB	GRASS	GROUND COVER

GROWTH (METRES)

H	2-3	W	1.5	SPACE	2

PRODUCTIVE SEASON

J	**F**	**M**	**A**	**M**	**J**	**J**	**A**	**S**	**O**	**N**	**D**

LIGHT REQUIREMENTS

SHADE	**PART SHADE**	FULL SUN

KELEDANG, *Artocarpus lanceifolius*

- Relative to jakfruit.
- Fruit is more tangy but sweet.
- Tree is more wide spread.
- Seeds can be eaten after cooking.
- Will grow tall so best kept pruned to access fruit and wind tolerance.
- Fruit within 3-4 years.

SHAPE			
TREE	PALM	SHRUB	CLUMP
VINE	HERB	GRASS	GROUND COVER

GROWTH (METRES)					
H	10-20	W	3	SPACE	3

PRODUCTIVE SEASON											
J	F	M	**A**	**M**	J	J	A	S	O	N	D

LIGHT REQUIREMENTS		
SHADE	PART SHADE	**FULL SUN**

LEMON ASPEN, *Acronychia acidula*

- Umbrella share tree
- Native to Australian rainforest at altitudes of 1000m.
- Fruits have a lemon / grapefruit flavour.
- Best kept pruned to access fruits.
- Very good source of folate, zinc and iron
- Prefers well drained fertile soils
- Fruit within 5 years

SHAPE			
TREE	PALM	SHRUB	CLUMP
VINE	HERB	GRASS	GROUND COVER

GROWTH (METRES)					
H	10	W	3	SPACE	3

PRODUCTIVE SEASON											
J	F	M	A	M	J	**J**	**A**	**S**	O	N	D

LIGHT REQUIREMENTS		
SHADE	**PART SHADE**	FULL SUN

LEMON, *Citrus × limon*

- Fruit used in various drinks and dishes
- Plant is best sourced as a grafted tree to produce within a few years.
- Likes a good dose of compost regularly
- Regular pruning needed to keep a small compact shape.
- If fruit is left on the tree the length of available fruit is extended.
- Meyer lemon doest very well in the tropics.

SHAPE			
TREE	PALM	SHRUB	CLUMP
VINE	HERB	GRASS	GROUND COVER

GROWTH (METRES)					
H	2	W	3	SPACE	3

PRODUCTIVE SEASON											
J	F	M	**A**	**M**	**J**	**J**	**A**	S	O	N	D

LIGHT REQUIREMENTS		
SHADE	**PART SHADE**	FULL SUN

LIME WEST INDIAN, *Citrus × aurantiifolia*

- A more juicy variety of lime
- Compact and high produce
- Keep pruned to maintain good shape
- High feeder so a good dose of compost is needed every 3 months
- Will grow from seed and be true to type
- Propagation by layering is recommended
- Fruit within 2-3 years

SHAPE			
TREE	PALM	**SHRUB**	CLUMP
VINE	HERB	GRASS	GROUND COVER

GROWTH (METRES)					
H	1.5	W	1.5	SPACE	2.5

PRODUCTIVE SEASON											
J	F	M	**A**	**M**	**J**	**J**	**A**	S	O	N	D

LIGHT REQUIREMENTS		
SHADE	**PART SHADE**	FULL SUN

LONGAN, Dimocarpus longan

- Large, strong and cold tolerant
- The fruit is deliciously sweet with more earthy tones
- A highly ornamental tree with beautiful green leaves and clusters of yellow flowers that have a wonderful scent
- Fruit within 7-8 years

SHAPE			
TREE	PALM	SHRUB	CLUMP
VINE	HERB	GRASS	GROUND COVER

GROWTH (METRES)					
H	8	W	5	SPACE	5

PRODUCTIVE SEASON											
J	**F**	**M**	A	M	J	J	A	S	O	N	D

LIGHT REQUIREMENTS		
SHADE	**PART SHADE**	**FULL SUN**

LOQUAT, *Eriobotrya japonica*

- Native to China.
- Can handle frosts, therefore more suitable to the sub-tropics.
- Evergreen.
- Likes neutral PH 6-7.
- Requires extra special care to fruit well in the tropics, provide composts and balanced minerals.
- Will fruit within 4 years.

SHAPE			
TREE	PALM	SHRUB	CLUMP
VINE	HERB	GRASS	GROUND COVER

GROWTH (METRES)					
H	4	W	3	SPACE	3

PRODUCTIVE SEASON											
J	F	M	A	M	J	J	A	**S**	**O**	**N**	D

LIGHT REQUIREMENTS		
SHADE	PART SHADE	**FULL SUN**

LYCHEE, *Litchi chinensis*

- Very juicy and sweet.
- Best kept pruned and cover with a net in fruiting season before the bats attack.
- Many different varieties and sizes with varied fruiting times.
- Cincturing or ring barking branches will help encourage flowering.
- Fruit within 9 years from seedling.

SHAPE			
TREE	PALM	SHRUB	CLUMP
VINE	HERB	GRASS	GROUND COVER

GROWTH (METRES)					
H	6-10	W	6	SPACE	4

PRODUCTIVE SEASON											
J	F	M	A	M	J	J	A	S	O	**N**	**D**

LIGHT REQUIREMENTS		
SHADE	PART SHADE	**FULL SUN**

MACADAMIA, *Macadamia integrifolia*

- As a Australian native Macadamia has evolved with little to no Phosphorus. This means that any fertilizer containing Phosphorus will harm the tree.
- Best grown from seed.
- Can be allowed to grow tall and harvest nuts will drop when ready.
- Fruit within 5 years.

SHAPE			
TREE	PALM	SHRUB	CLUMP
VINE	HERB	GRASS	GROUND COVER

GROWTH (METRES)					
H	8	W	5	SPACE	5

PRODUCTIVE SEASON											
J	**F**	**M**	**A**	**M**	**J**	**J**	**A**	**S**	O	N	D

LIGHT REQUIREMENTS		
SHADE	PART SHADE	**FULL SUN**

MALABAR CHESTNUT, *Pachira glabra*

- The seeds can be eaten either raw or roasted and taste like peanut / coconut
- The young leaves and flowers are cooked and eaten as a vegetable
- Flowers are very attractive, large, white and fragrant
- Can grow in clay with poor drainage
- Fruit within 3-4 years

SHAPE			
TREE	PALM	SHRUB	CLUMP
VINE	HERB	GRASS	GROUND COVER

GROWTH (METRES)					
H	5	W	4	SPACE	4

PRODUCTIVE SEASON											
J	F	M	A	M	J	J	**A**	**S**	**O**	N	D

LIGHT REQUIREMENTS		
SHADE	**PART SHADE**	**FULL SUN**

MAMEY SAPOTE, *Pouteria sapota*

- Fruit is large with orange flesh tasting like sweet roast potato/ pumpkin
- Can be eaten fresh or made into cakes or smoothies
- Fruit will take many months to ripen on the tree, just scratch the skin a little and if orange it's ready to pick
- Large tree so keep pruned to size
- Fruit within 5-6 years

SHAPE			
TREE	PALM	SHRUB	CLUMP
VINE	HERB	GRASS	GROUND COVER

GROWTH (METRES)					
H	10	W	8	SPACE	5

PRODUCTIVE SEASON											
J	F	**M**	**A**	**M**	**J**	J	A	S	O	N	D

LIGHT REQUIREMENTS		
SHADE	PART SHADE	**FULL SUN**

MANGO PLUM, *Bouea macrophylla*

- Edible skin with sweet crispy flesh that tastes like an apricot mango
- Can be a large tree so best kept pruned
- Immature fruits can be made into chutney
- Young leaves can be used in salad
- Fruit within 5 years from seedling

SHAPE			
TREE	PALM	SHRUB	CLUMP
VINE	HERB	GRASS	GROUND COVER

GROWTH (METRES)					
H	6-10	W	6	SPACE	4

PRODUCTIVE SEASON											
J	**F**	**M**	A	M	J	J	A	S	O	N	D

LIGHT REQUIREMENTS		
SHADE	PART SHADE	**FULL SUN**

MANGO, *Mangifera indica*

- Mangoes have either mono or polyembryonic, polyembryonic seeds grow true to type, which are Kensington Pride, R2E2 & Nom Doc Mai
- Keitt is a late-season, best suited to sub tropics
- Keep tree pruned to size
- Fruit within 4-5 years from seed

SHAPE			
TREE	PALM	SHRUB	CLUMP
VINE	HERB	GRASS	GROUND COVER

GROWTH (METRES)					
H	3.5	W	3	SPACE	3

PRODUCTIVE SEASON											
J	**F**	**M**	**A**	M	J	J	A	S	O	N	D

LIGHT REQUIREMENTS		
SHADE	PART SHADE	**FULL SUN**

MANGOSTEEN, Garcinia mangostana

- A slow-growing tree with a pyramidal crown that is native to rainforests of Malaysia and Indonesia.
- Considered the queen of fruit.
- Can grow in the under-story of other large trees.
- Grown from seed.
- Fruit within 10 years.

SHAPE			
TREE	PALM	**SHRUB**	CLUMP
VINE	HERB	GRASS	GROUND COVER

GROWTH (METRES)					
H	3.5	W	3	SPACE	3

PRODUCTIVE SEASON											
J	**F**	**M**	**A**	M	J	J	A	S	O	N	D

LIGHT REQUIREMENTS		
SHADE	**PART SHADE**	FULL SUN

MARANG, Artocarpus odoratissimus

- Resembles both the jackfruit and the breadfruit in appearance
- The fruit is succulent and mildly flavoured, regarded as superior to both jackfruit and chempedak
- Protect from wind as the branches can easily break
- Two plants for pollination
- Fruit within 5-7 years

SHAPE			
TREE	PALM	SHRUB	CLUMP
VINE	HERB	GRASS	GROUND COVER

GROWTH (METRES)					
H	10+	W	7	SPACE	4

PRODUCTIVE SEASON											
J	**F**	**M**	**A**	**M**	**J**	J	A	S	O	N	D

LIGHT REQUIREMENTS		
SHADE	PART SHADE	**FULL SUN**

MIRACLE FRUIT, *Synsepalum dulcificum*

- Small red fruit that makes everything taste sweet for up too an hour after eating them.
- Slender shrub that grows tall under the shade of other trees.
- Will grow tall and thin if left un-pruned.
- Makes a great hedge with multiple plants.
- Fruit within 2-3 years from seed.

SHAPE			
TREE	PALM	**SHRUB**	CLUMP
VINE	HERB	GRASS	GROUND COVER

GROWTH (METRES)					
H	4	W	1	SPACE	1

PRODUCTIVE SEASON											
J	F	M	A	**M**	**J**	**J**	**A**	**S**	**O**	N	D

LIGHT REQUIREMENTS		
SHADE	PART SHADE	FULL SUN

MONSTERIA, *Monstera deliciosa*

- Low growing but can climb trees, best kept to the ground to pick fruits.
- Attractive leaf structure.
- Fruit tastes like fruit salad, pineapple, mango, banana.
- Each morsel of the fruit can only be eaten as the skin falls off as it can be irritating.
- Fruit within 1-2 years.

SHAPE

TREE	PALM	**SHRUB**	CLUMP
VINE	HERB	GRASS	GROUND COVER

GROWTH (METRES)

H		W		SPACE	
H	2.5	W	2	SPACE	1

PRODUCTIVE SEASON

J	**F**	**M**	**A**	**M**	**J**	J	A	S	O	N	D

LIGHT REQUIREMENTS

SHADE	**PART SHADE**	FULL SUN

PANAMA BERRY, *Muntingia calabura*

- Ornamental tree
- Flowers and fruits most of the year
- Long weeping branches
- Delicious fruit tastes like caramel
- Can be pruned to shape
- Fast growing
- Fruit within 2 years

SHAPE

TREE	PALM	SHRUB	CLUMP
VINE	HERB	GRASS	GROUND COVER

GROWTH (METRES)

H	5	W	5	SPACE	4

PRODUCTIVE SEASON

J	**F**	**M**	**A**	M	J	**J**	**A**	**S**	**O**	**N**	**D**

LIGHT REQUIREMENTS

SHADE	PART SHADE	**FULL SUN**

PAPAYA, *Carica Papaya*

- Fruit and vegetable.
- Can be affected by scale bug, an application of white oil can help with this.
- Medicinal leaves.
- Replace or prune when they get high.
- Stems are hollow and used as straws.
- They like a good dose of calcium and lime for optimum fruiting.
- Fruit within 1 year.

SHAPE

TREE	PALM	SHRUB	CLUMP
VINE	HERB	GRASS	GROUND COVER

GROWTH (METRES)

H	3-5	W	1.5	SPACE	2.5

PRODUCTIVE SEASON

J	**F**	**M**	**A**	**M**	**J**	**J**	**A**	**S**	**O**	**N**	**D**

LIGHT REQUIREMENTS

SHADE	**PART SHADE**	FULL SUN

PASSIONFRUIT, *Passiflora edulis*

- Productive vine with delicious fruit.
- Requires trellising.
- After first year prune growth tip to help branching.
- Prune back old growth in early spring.
- Vines can be easily grown from the seeds within the fruit.
- Many types; purple, panama gold, banana, native.

SHAPE

TREE	PALM	SHRUB	CLUMP
VINE	HERB	GRASS	GROUND COVER

GROWTH (METRES)

H	5	W	5	SPACE	2.5

PRODUCTIVE SEASON

J	**F**	**M**	**A**	M	J	J	A	S	O	**N**	**D**

LIGHT REQUIREMENTS

SHADE	**PART SHADE**	FULL SUN

PEANUT BUTTER FRUIT, *Bunchosia argentea*

- Attractive small tree/ shrub
- Ripe fruit taste of a raspberry/date with the texture peanut butter when fully ripe (red).
- Crispy carrot texture when under ripe (orange).
- Can be grown and pruned as a hedge.
- Fruit within 2-3 years.

SHAPE

TREE	PALM	**SHRUB**	CLUMP
VINE	HERB	GRASS	GROUND COVER

GROWTH (METRES)

H	3	W	3	SPACE	2

PRODUCTIVE SEASON

J	**F**	**M**	**A**	M	J	J	A	S	O	N	D

LIGHT REQUIREMENTS

SHADE	**PART SHADE**	FULL SUN

PEPINO, Solanum muricatum

- Small bush ground cover with small fruit that taste like rockmelon.
- Fruit ready to pick when slightly soft and fragrant.
- Prefer plenty of water in fertile soils
- Grows from cuttings.
- Relative to tomato so can be susceptible to similar diseases and insects.

SHAPE			
TREE	PALM	**SHRUB**	CLUMP
VINE	HERB	GRASS	**GROUND COVER**

GROWTH (METRES)					
H	.5	W	1.5	SPACE	2.5

PRODUCTIVE SEASON											
J	F	**M**	**A**	**M**	**J**	**J**	**A**	S	O	N	D

LIGHT REQUIREMENTS		
SHADE	**PART SHADE**	FULL SUN

PEPPER, *Piper nigrum*

- Valuable and highly used spice.
- Harvest seeds when they are red.
- Vines reduce in production after 5 years so keep propagating new plants.
- Grown from seed or cutting.
- Need a timber post or trellis to grow up.
- Fruit within 2 years.

SHAPE			
TREE	PALM	SHRUB	CLUMP
VINE	HERB	GRASS	GROUND COVER

GROWTH (METRES)					
H	3-5	W	1.5	SPACE	2.5

PRODUCTIVE SEASON											
J	**F**	**M**	**A**	**M**	**J**	J	A	S	O	N	D

LIGHT REQUIREMENTS		
SHADE	**PART SHADE**	FULL SUN

PERSIMMON, *Diospyros kaki*

- The two fruiting types are made up of astringent and non astringent.
- Nice and sweet and good market value fruit .
- A very attractive deciduous tree with spectacular dark-orange autumn foliage.
- Suited better to the dry sub-tropics.
- Fruit within 4-5 years when grown from seed.

SHAPE			
TREE	PALM	SHRUB	CLUMP
VINE	HERB	GRASS	GROUND COVER

GROWTH (METRES)					
H	3	W	3	SPACE	3.5

PRODUCTIVE SEASON											
J	F	**M**	**A**	**M**	**J**	**J**	A	S	O	N	D

LIGHT REQUIREMENTS		
SHADE	**PART SHADE**	FULL SUN

PINEAPPLE, *Ananas comosus*

- Good edging to deter animals
- Spineless varieties are preferred for ease of garden maintenance
- Prefer slightly acid soils
- Can be easily grown from planting the top, but will fruit quicker when grown from a pup (less than a year)

SHAPE			
TREE	PALM	SHRUB	**CLUMP**
VINE	HERB	GRASS	GROUND COVER

GROWTH (METRES)					
H	.75	W	.75	SPACE	.75

PRODUCTIVE SEASON											
J	**F**	**M**	**A**	M	J	J	A	S	O	**N**	**D**

LIGHT REQUIREMENTS		
SHADE	PART SHADE	FULL SUN

POMEGRANATE, *Punica granatum*

- Produces very large fruit to 1.5kg that is very sweet and juicy. Very adaptable and will grow in regions ranging from temperate to tropical
- Provide adequate water at first flush of flowers
- Deciduous and drought-tolerant
- Grown from cutting
- Fruit within 2-3 years

SHAPE			
TREE	PALM	SHRUB	CLUMP
VINE	HERB	GRASS	GROUND COVER

GROWTH (METRES)					
H	3	W	3.5	SPACE	3

PRODUCTIVE SEASON											
J	F	**M**	**A**	**M**	J	J	A	S	O	N	D

LIGHT REQUIREMENTS		
SHADE	PART SHADE	**FULL SUN**

RAMBUTAN, *Nephelium lappaceum*

- Male and female trees required for fruit, so recommended minimum of 3 plants to each garden.
- Best kept pruned to access fruit.
- Super tropical, so will only grow in coastal wet tropics.
- Fruit within 5-6 years.

SHAPE			
TREE	PALM	SHRUB	CLUMP
VINE	HERB	GRASS	GROUND COVER

GROWTH (METRES)					
H	7-10	W	3	SPACE	2

PRODUCTIVE SEASON											
J	**F**	**M**	A	M	J	J	A	S	O	N	D

LIGHT REQUIREMENTS		
SHADE	PART SHADE	**FULL SUN**

ROLLINIA, *Rollinia deliciosa*

- The fruit has a creamy juicy and delicious flesh, according to some tasting like lemon meringue pie
- Similar to custard apple but with creamier and less sweet taste
- Tendency to grow tall and wide
- Will fruit twice a year
- Fruits within 4 years from seed

SHAPE			
TREE	PALM	SHRUB	CLUMP
VINE	HERB	GRASS	GROUND COVER

GROWTH (METRES)					
H	5	W	7	SPACE	7

PRODUCTIVE SEASON											
J	F	**M**	**A**	**M**	**J**	J	A	S	**O**	**N**	D

LIGHT REQUIREMENTS		
SHADE	PART SHADE	**FULL SUN**

SAPODILLA, *Manilkara zapota*

- Hardy tree and highly drought resistant when mature.
- Young trees require irrigation.
- Can withstand some salinity.
- The latex obtained from the bark of the tree can be used as a chewing gum.
- Two plants required for pollination.
- Fruit within 5 years from seed.

SHAPE			
TREE	PALM	SHRUB	CLUMP
VINE	HERB	GRASS	GROUND COVER

GROWTH (METRES)					
H	4	W	4	SPACE	4

PRODUCTIVE SEASON											
J	**F**	**M**	A	M	J	J	A	S	O	N	D

LIGHT REQUIREMENTS		
SHADE	PART SHADE	**FULL SUN**

SOURSOP, *Annona muricata*

- The Soursop is native to the northern parts of South America.
- The fruit is heart-shaped with a rough green skin and soft fleshy spines.
- The fresh meaty flesh is juicy and slightly acid, producing a rich creamy thirst quenching juice.
- Fruit within 2-3 years from seed.

SHAPE			
TREE	PALM	SHRUB	CLUMP
VINE	HERB	GRASS	GROUND COVER

GROWTH (METRES)					
H	4	W	4	SPACE	4

PRODUCTIVE SEASON											
J	**F**	**M**	**A**	**M**	**J**	J	A	S	O	N	D

LIGHT REQUIREMENTS		
SHADE	PART SHADE	**FULL SUN**

STAR APPLE, *Chrysophyllum cainito*

- A spectacular ornamental tree that has large green leaves with a vibrant golden underside.
- Fruit is sweet and juicy with a wonderful caramel vanilla milkshake flavour.
- Keep pruned to maintain a manageable size.
- Fruit within 5 years from seed.

SHAPE			
TREE	PALM	SHRUB	CLUMP
VINE	HERB	GRASS	GROUND COVER

GROWTH (METRES)					
H	15	W	4	SPACE	4

PRODUCTIVE SEASON											
J	F	M	A	M	J	J	A	S	O	**N**	**D**

LIGHT REQUIREMENTS		
SHADE	PART SHADE	**FULL SUN**

STRAWBERRY, *Fragaria × ananassa*

- Perennial ground cover
- Propagate by runners
- To prevent fruit rot and disease remove dead leaves around plants and mulch with pine bark or needles
- If aphids appear control immediately with soapy water or pyrethrum
- Feed regularly with seaweed solution

SHAPE			
TREE	PALM	SHRUB	CLUMP
VINE	HERB	GRASS	**GROUND COVER**

GROWTH (METRES)					
H	.5	W	1.5	SPACE	2.5

PRODUCTIVE SEASON											
J	**F**	M	A	M	J	J	A	**S**	**O**	**N**	**D**

LIGHT REQUIREMENTS		
SHADE	PART SHADE	FULL SUN

TAMARILLO, *Cyphomandra betacea*

- Small shrub
- Delicate softwood
- Prefers shade
- Provides delicious fruit
- Can be pruned to shape
- Fast growing and short lived
- Will fruit within 2-3 years

SHAPE			
TREE	PALM	**SHRUB**	CLUMP
VINE	HERB	GRASS	GROUND COVER

GROWTH (METRES)					
H	3.5	W	2	SPACE	2

PRODUCTIVE SEASON											
J	F	M	A	M	J	**J**	**A**	**S**	**O**	**N**	**D**

LIGHT REQUIREMENTS		
SHADE	**PART SHADE**	FULL SUN

WEST INDIAN LOCUST, *Hymenaea courbaril*

- Large tree
- Produces an abundant crop after 8 years
- Tastes like caramel malt powder
- Used in treats and smoothies
- Fruit eaten fresh
- High in protein
- Leaves are cooked in soup
- Fruit can be stored and last most the year

SHAPE			
TREE	PALM	SHRUB	CLUMP
VINE	HERB	GRASS	GROUND COVER

GROWTH (METRES)					
H	25	W	15	SPACE	15

PRODUCTIVE SEASON											
J	**F**	M	A	M	J	J	A	S	O	**N**	**D**

LIGHT REQUIREMENTS		
SHADE	PART SHADE	**FULL SUN**

WHITE SAPOTE, *Casimiroa edulis*

- Native to the Mexican highlands and Central America
- Fruit tasting reminiscent of vanilla custard mixed with creamy banana and peaches
- Roots can spread and damage houses
- Two plants are beneficial for bigger crops
- Fruit within 5-6 years from seed

SHAPE			
TREE	PALM	SHRUB	CLUMP
VINE	HERB	GRASS	GROUND COVER

GROWTH (METRES)					
H	10	W	4	SPACE	4

PRODUCTIVE SEASON											
J	**F**	**M**	A	M	J	J	A	S	O	N	D

LIGHT REQUIREMENTS		
SHADE	PART SHADE	**FULL SUN**

WHITE SHAHTOOT, *Morus alba*

- Large, sweet fruit can reach up to 10cm in length.
- Delicious eaten fresh but can dried.
- Spreading tree with a weeping habit, excellent shade tree.
- You can also get multiple crops by pruning immediately after your first crop.
- Propagation by heal cutting or marcott
- Fruit within 2-3 years

SHAPE			
TREE	PALM	SHRUB	CLUMP
VINE	HERB	GRASS	GROUND COVER

GROWTH (METRES)					
H	6	W	4	SPACE	4

PRODUCTIVE SEASON											
J	F	M	A	M	J	J	A	S	**O**	**N**	D

LIGHT REQUIREMENTS		
SHADE	PART SHADE	**FULL SUN**

BUNYA PINE, *Arauracaria bidwilli*

- Evergreen conifer with sharp, triangular leaves and a tall, symmetrical shape.
- Native to north and south QLD Australia.
- Massive cones weighing up to 10 kg, each containing dozens of nuts.
- Nuts are edible after roasting, be sure to split husk first to avoid an explosion.
- Fruit after 10 years and then better every second year.

SHAPE			
TREE	PALM	SHRUB	CLUMP
VINE	HERB	GRASS	GROUND COVER

GROWTH (METRES)					
H	20	W	10	SPACE	15

PRODUCTIVE SEASON											
J	**F**	M	A	M	J	J	A	S	O	N	D

LIGHT REQUIREMENTS		
SHADE	PART SHADE	**FULL SUN**

BRAZILIAN CHERRY, Eugenia uniflora

- Produces small, ribbed, bright red fruits with a sweet-tart flavor, often eaten fresh or used in jams.
- Rich in vitamin C and antioxidants, it supports immunity and overall health.
- The plant is an evergreen shrub with glossy leaves and fragrant white flowers.
- Grown as a dense shrub fro screening.

SHAPE			
TREE	PALM	**SHRUB**	CLUMP
VINE	HERB	GRASS	GROUND COVER

GROWTH (METRES)					
H	2	W	1.5	SPACE	.5

PRODUCTIVE SEASON											
J	F	M	A	M	J	J	A	**S**	**O**	**N**	D

LIGHT REQUIREMENTS		
SHADE	**PART SHADE**	FULL SUN

KOLA, *Cola acuminata*

- Chewed fresh for its caffeine-rich, stimulant properties in West Africa.
- The nuts are typically white or reddish in color and are known to boost energy and reduce fatigue.
- Historically flavoured cola beverages with its bitter, nutty taste.
- Tree is evergreen, glossy leaves and star-shaped fruit pods containing multiple nuts.

SHAPE			
TREE	PALM	SHRUB	CLUMP
VINE	HERB	GRASS	GROUND COVER

GROWTH (METRES)					
H	5	W	4	SPACE	4

PRODUCTIVE SEASON											
J	**F**	M	A	M	J	J	A	S	O	N	D

LIGHT REQUIREMENTS		
SHADE	**PART SHADE**	**FULL SUN**

MADRONO, *Garcinia intermedia*

- Similar to Achacha but smaller and slightly more acidic taste yet still yummy.
- Requires more than one plant to fruit well although one plant will still bare.
- Can handle little sunlight, so can be ground in dense under story.
- Fruit within 3-5 years.
- Perfect for subburban backyards.

SHAPE			
TREE	PALM	SHRUB	CLUMP
VINE	HERB	GRASS	GROUND COVER

GROWTH (METRES)					
H	3	W	3	SPACE	2

PRODUCTIVE SEASON											
J	**F**	**M**	A	M	J	J	A	S	O	N	D

LIGHT REQUIREMENTS		
SHADE	PART SHADE	FULL SUN

RED SHAHTOOT, Morus macroura

- Large, sweet fruit can reach up to 10cm in length.
- Spreading tree with a weeping habit, excellent shade tree
- You can also get multiple crops by pruning immediately after your first crop
- Propagation by heal cutting or marcott
- Fruit within 2-3 years

SHAPE			
TREE	PALM	SHRUB	CLUMP
VINE	HERB	GRASS	GROUND COVER

GROWTH (METRES)					
H	6	W	4	SPACE	4

PRODUCTIVE SEASON											
J	F	M	A	M	J	J	A	S	**O**	**N**	D

LIGHT REQUIREMENTS		
SHADE	**PART SHADE**	**FULL SUN**

AIBIKA, *Abelmoschus manihot*

- This plant is a short-lived perennial vegetable with many different varieties.
- Best kept to about 1m tall as this can grow to about 3-4m in height and width.
- Replant as a direct cutting into the ground throughout the wet season.
- Leaves are used as spinach and the tips when cooked are like broccoli.
- Will turn a soup slimy if over cooked.

SHAPE			
TREE	PALM	**SHRUB**	CLUMP
VINE	HERB	GRASS	GROUND COVER

GROWTH (METRES)					
H	3	W	3	SPACE	1

PRODUCTIVE SEASON											
J	**F**	**M**	**A**	**M**	**J**	J	A	**S**	**O**	**N**	**D**

LIGHT REQUIREMENTS		
SHADE	**PART SHADE**	FULL SUN

ASPARAGUS, *Asparagus officinalis*

- Best planted in a group of at least 6 to ensure substantial crop.
- Allow 1 year before harvesting any spears to ensure good root development.
- Potential weed in sub-tropics.

SHAPE			
TREE	PALM	SHRUB	**CLUMP**
VINE	HERB	GRASS	GROUND COVER

GROWTH (METRES)					
H	1	W	1	SPACE	1

PRODUCTIVE SEASON											
J	**F**	**M**	A	M	J	J	A	S	**O**	**N**	**D**

LIGHT REQUIREMENTS		
SHADE	**PART SHADE**	FULL SUN

BAMBOO, *Bambusa sp.*

- Clumping or running bamboo can be eaten after cooking or even raw
- Can be used in construction, craft and tool making
- Best to plant clumping species as running varieties are hard to maintain
- See appendix for table of clumping species on page 28

SHAPE			
TREE	PALM	SHRUB	CLUMP
VINE	HERB	**GRASS**	GROUND COVER

GROWTH (METRES)					
H	10	W	4.5	SPACE	4

PRODUCTIVE SEASON											
J	**F**	**M**	**A**	**M**	**J**	J	A	S	O	N	D

LIGHT REQUIREMENTS		
SHADE	PART SHADE	**FULL SUN**

BAY LAUREL, *Laurus nobilis*

- Hardy evergreen tree with deep green aromatic foliage.
- Leaves are used to flavour soups & stews.
- Keep pruned to a bush to encourage new growth.
- Requires good drainage.
- Successfully grown in the wet tropics for many years on the top of a rock herb spiral.

SHAPE			
TREE	PALM	**SHRUB**	CLUMP
VINE	HERB	GRASS	GROUND COVER

GROWTH (METRES)					
H	1.5	W	1	SPACE	.5

PRODUCTIVE SEASON											
J	**F**	**M**	**A**	**M**	**J**	**J**	**A**	**S**	**O**	**N**	**D**

LIGHT REQUIREMENTS		
SHADE	**PART SHADE**	**FULL SUN**

BETEL LEAF, *Piper sarmentosum*

- Used to wrap foods and can be eaten fresh or cooked.
- Try soaking in sugar water for couple hours before serving to improve taste.
- Very vigorous ground cover that grows in dense shade under trees or in full sun.
- Difficult to remove once established and can become weedy, keep away from veggie gardens.

SHAPE			
TREE	PALM	SHRUB	CLUMP
VINE	HERB	GRASS	**GROUND COVER**

GROWTH (METRES)					
H	.5	W	-	SPACE	-

PRODUCTIVE SEASON											
J	**F**	**M**	**A**	**M**	**J**	**J**	**A**	**S**	**O**	**N**	**D**

LIGHT REQUIREMENTS		
SHADE	PART SHADE	FULL SUN

BRAHMI, *Bacopa monnieri*

- Water loving ground cover herb.
- By eating a little each day is said to be used to enhance memory and brain function.
- Delicate small light green leaves.
- Grows well in the shade.

SHAPE			
TREE	PALM	SHRUB	CLUMP
VINE	**HERB**	GRASS	**GROUND COVER**

GROWTH (METRES)					
H	.1	W	-	SPACE	-

PRODUCTIVE SEASON											
J	**F**	**M**	**A**	**M**	**J**	**J**	**A**	**S**	**O**	**N**	**D**

LIGHT REQUIREMENTS		
SHADE	PART SHADE	FULL SUN

BRAZILIAN SPINACH, *Alternanthera sissoo*

- Perennial spinach.
- When grown in shade the leaves are more tender.
- Plant around garden edges to hold soil and give form.
- Excellent chicken food, great in grazing frames.
- Two known types, round and long leaf (pictured).

SHAPE			
TREE	PALM	SHRUB	CLUMP
VINE	**HERB**	GRASS	**GROUND COVER**

GROWTH (METRES)					
H	.2	W	1	SPACE	-

PRODUCTIVE SEASON											
J	**F**	**M**	**A**	**M**	**J**	**J**	**A**	**S**	**O**	**N**	**D**

LIGHT REQUIREMENTS		
SHADE	PART SHADE	FULL SUN

CASSAVA, *Manihot esculenta*

- Used as a potato when boiled and leaves can be eaten after cooking.
- Cassava is one of the highest yielding carbohydrate sources.
- Replant as a direct cutting into the ground through the wet season.
- Must be eaten within 2 days of harvest as cyanide will build up.

SHAPE			
TREE	PALM	**SHRUB**	CLUMP
VINE	HERB	GRASS	GROUND COVER

GROWTH (METRES)					
H	3	W	1	SPACE	.75

PRODUCTIVE SEASON											
J	F	M	**A**	**M**	**J**	**J**	**A**	**S**	**O**	N	D

LIGHT REQUIREMENTS		
SHADE	**PART SHADE**	**FULL SUN**

CELERY TARO, *Colocasia esculenta*

- A variety of taro.
- Stems can be eaten in soups and stir-fry.
- The main corm is not a palatable to commercial taro.
- Propagation is done by dividing the little cormlets at the base of the plant and replanting during the wet season.
- Doesn't need very wet soil.

SHAPE			
TREE	PALM	SHRUB	**CLUMP**
VINE	HERB	GRASS	GROUND COVER

GROWTH (METRES)					
H	2	W	1	SPACE	1

PRODUCTIVE SEASON											
J	**F**	**M**	**A**	**M**	**J**	**J**	**A**	**S**	**O**	**N**	**D**

LIGHT REQUIREMENTS		
SHADE	**PART SHADE**	FULL SUN

CHILLI, *Capsicum annuum*

- There are over 200 varieties of chilli and all have varying degrees of spice.
- Hardy perennial.
- A spray can be made to deter caterpillars on other crops.
- Propagation from seed so seed save for strong chilli stock for future seasons.
- Produces fruit of many colours including red, green, yellow, purple and striped.

SHAPE			
TREE	PALM	SHRUB	**CLUMP**
VINE	HERB	GRASS	GROUND COVER

GROWTH (METRES)					
H	2	W	1	SPACE	1

PRODUCTIVE SEASON											
J	**F**	**M**	A	M	J	J	A	**S**	**O**	**N**	**D**

LIGHT REQUIREMENTS		
SHADE	**PART SHADE**	**FULL SUN**

CHIVES, *Tulbaghia violacea*

- A tight clumping perennial with attractive purple flower.
- Leaves have a strong garlic smell and flavour.
- Use as a boarder around gardens.
- Rich on vitamin C.
- Drought tolerant.

SHAPE			
TREE	PALM	SHRUB	**CLUMP**
VINE	HERB	GRASS	GROUND COVER

GROWTH (METRES)					
H	.4	W	.05	SPACE	.05

PRODUCTIVE SEASON											
J	**F**	**M**	**A**	**M**	**J**	**J**	**A**	**S**	**O**	**N**	**D**

LIGHT REQUIREMENTS		
SHADE	**PART SHADE**	FULL SUN

CHOKO, *Sicyos edule*

- A hardy summer perennial vine.
- Fruit and fresh vine tips are edible raw or cooked and are delicious.
- The roots are also highly susceptible to rot from excessive water logging.
- Relative to the gourd family.
- Propagate from the entire fruit.
- Harvest fruit before skin becomes tough.

SHAPE			
TREE	PALM	SHRUB	CLUMP
VINE	HERB	GRASS	GROUND COVER

GROWTH (METRES)					
H	10	W	10	SPACE	1

PRODUCTIVE SEASON											
J	**F**	**M**	**A**	M	J	J	A	**S**	**O**	**N**	**D**

LIGHT REQUIREMENTS		
SHADE	**PART SHADE**	FULL SUN

COCOYAM, *Xanthansoma sagittifolium*

- A variety of taro.
- The entire plant is large in comparison to most other taros.
- Dig around the base of an established Cocoyam to find a mine of tasty potato like taro.
- The cormlets are eaten instead of the main corm which is high in oxalic acids.
- Likes a lot of moisture in the soil.

SHAPE			
TREE	PALM	SHRUB	**CLUMP**
VINE	HERB	GRASS	GROUND COVER

GROWTH (METRES)					
H	2	W	2.5	SPACE	1.5

PRODUCTIVE SEASON											
J	F	M	A	**M**	**J**	**J**	**A**	**S**	O	N	D

LIGHT REQUIREMENTS		
SHADE	**PART SHADE**	FULL SUN

COMFREY, *Symphytum officinale*

- Compost activator and soil builder.
- Contains allantoin, a substance that helps new skin cells grow.
- Used medically to heal minor wounds, sprains, bruises and broken bones.
- Propagation by root division.
- Have know people to batter and deep fry for a snack.

SHAPE			
TREE	PALM	SHRUB	**CLUMP**
VINE	**HERB**	GRASS	GROUND COVER

GROWTH (METRES)					
H	1	W	1	SPACE	.5

PRODUCTIVE SEASON											
J	**F**	**M**	**A**	**M**	**J**	**J**	**A**	**S**	**O**	**N**	**D**

LIGHT REQUIREMENTS		
SHADE	**PART SHADE**	FULL SUN

CRANBERRY HIBISCUS, *Hibiscus acetosella*

- Perennial shrub the self seeds readily.
- Zesty flavoured leaf is great in salads.
- Has a tendency to get leggy, so best to keep pruned.
- Deep burgundy coloured plant.
- Grown from seed or cutting.
- It will form a calyx which is not eaten like is relative Rosella.

SHAPE			
TREE	PALM	**SHRUB**	CLUMP
VINE	HERB	GRASS	GROUND COVER

GROWTH (METRES)					
H	2	W	1	SPACE	1

PRODUCTIVE SEASON											
J	**F**	**M**	**A**	**M**	**J**	**J**	**A**	**S**	**O**	**N**	**D**

LIGHT REQUIREMENTS		
SHADE	**PART SHADE**	**FULL SUN**

CURRY TREE, *Murraya koenigii*

- Leaves used to flavour food for curry.
- Invasive with its lateral roots, keep in a pot in urban areas.
- Evergreen.
- Propagated by root division or seedling.
- Requires moderate watering.
- Fruits are edible and nutritious.
- Keep pruned as a shrub.

SHAPE			
TREE	PALM	SHRUB	CLUMP
VINE	HERB	GRASS	GROUND COVER

GROWTH (METRES)					
H	3	W	3	SPACE	2

PRODUCTIVE SEASON											
J	**F**	**M**	**A**	**M**	**J**	**J**	**A**	**S**	**O**	**N**	**D**

LIGHT REQUIREMENTS		
SHADE	**PART SHADE**	**FULL SUN**

EGGPLANT, *Solanum melongena*

- Many varieties of shapes, colours and sizes
- With good drainage an eggplant can last for a couple of years especially when grafted onto a wild root stock
- Drought tolerant
- Grown from seed

SHAPE			
TREE	PALM	**SHRUB**	CLUMP
VINE	HERB	GRASS	GROUND COVER

GROWTH (METRES)					
H	1	W	2	SPACE	1.5

PRODUCTIVE SEASON											
J	**F**	**M**	**A**	**M**	J	J	A	S	O	**N**	**D**

LIGHT REQUIREMENTS		
SHADE	PART SHADE	**FULL SUN**

GALANGAL, *Alpinia galanga*

- Large dense clumping perennial.
- Produces fragrant rhizome roots used as a spice like ginger.
- Good for screening.
- Propagated from root division.

SHAPE			
TREE	PALM	SHRUB	**CLUMP**
VINE	HERB	GRASS	GROUND COVER

GROWTH (METRES)					
H	2.5	W	2	SPACE	1

PRODUCTIVE SEASON											
J	**F**	**M**	**A**	**M**	**J**	**J**	**A**	**S**	**O**	**N**	**D**

LIGHT REQUIREMENTS		
SHADE	**PART SHADE**	**FULL SUN**

GINGER, *Zingiber officinale*

- Edible tubers, for fragrant kitchen spice
- Young rhizomes are juicy and fleshy with a very mild taste.
- Steep in boiling water to make ginger tea
- Valuable crop on the market.
- Best grown from a large piece of root and mulched thickly with hay.

SHAPE			
TREE	PALM	SHRUB	**CLUMP**
VINE	**HERB**	GRASS	GROUND COVER

GROWTH (METRES)					
H	1	W	1	SPACE	1

PRODUCTIVE SEASON											
J	F	M	A	M	J	**J**	**A**	**S**	**O**	**N**	D

LIGHT REQUIREMENTS		
SHADE	**PART SHADE**	FULL SUN

GOTU KOLA, *Centella asiatica*

- Mild, slightly bitter leaves eaten fresh or used in salads, teas, and traditional medicine, however only eat a few a day.
- It grows as a low-spreading, creeping herb in moist, shaded areas.
- Propagates easily through runners, forming dense ground cover.
- Known for boosting brain function, wound healing, and circulation.

SHAPE			
TREE	PALM	SHRUB	CLUMP
VINE	**HERB**	GRASS	**GROUND COVER**

GROWTH (METRES)					
H	.5	W	.5	SPACE	.2

PRODUCTIVE SEASON											
J	**F**	**M**	**A**	**M**	**J**	**J**	**A**	**S**	**O**	**N**	**D**

LIGHT REQUIREMENTS		
SHADE	PART SHADE	FULL SUN

JICAMA, *Pachyrhizus erosus*

- It is a crunchy, root vegetable often eaten raw in salads.
- The plant produces vines and pods, but only the root is edible; other parts are toxic.
- Has a similar texture to unripe pear
- Vigorous vine that is typically annual but self seeds every year.

SHAPE			
TREE	PALM	SHRUB	CLUMP
VINE	HERB	GRASS	GROUND COVER

GROWTH (METRES)					
H	-	W	.5	SPACE	.2

PRODUCTIVE SEASON											
J	F	M	A	**M**	**J**	**J**	**A**	S	O	N	D

LIGHT REQUIREMENTS		
SHADE	**PART SHADE**	FULL SUN

LEAF GINSENG, *Talinum triangulare*

- Perennial herb that is used as a spinach or salad green.
- Small pink flower is good bee fodder.
- Leaves are tangy and crunchy.
- Similar medicinally to Panax Ginseng which is used for blood building and rejuvenating tonics.
- Self propagates from seed or from cutting.

SHAPE			
TREE	PALM	**SHRUB**	CLUMP
VINE	**HERB**	GRASS	GROUND COVER

GROWTH (METRES)					
H	.5	W	.5	SPACE	.2

PRODUCTIVE SEASON											
J	**F**	**M**	**A**	**M**	**J**	**J**	**A**	**S**	**O**	**N**	**D**

LIGHT REQUIREMENTS		
SHADE	**PART SHADE**	**FULL SUN**

LEBANESE CRESS, *Aethionema cordifolium*

- Perennial herb that is used as a salad green with fresh flavour similar to carrot.
- Has a creeping nature and will spread throughout the garden.
- Grows in water and in free draining soil.
- Tends to die back during summer with heat so keep it in a cool shady spot.

SHAPE			
TREE	PALM	SHRUB	CLUMP
VINE	**HERB**	GRASS	GROUND COVER

GROWTH (METRES)					
H	.1	W	.1	SPACE	.1

PRODUCTIVE SEASON											
J	F	**M**	**A**	**M**	**J**	**J**	**A**	**S**	**O**	N	D

LIGHT REQUIREMENTS		
SHADE	PART SHADE	FULL SUN

LEMON GRASS, *Cymbopogon schoenanthus*

- A hardy perennial grass, not be confused with citronella which has a red stem.
- To encourage new growth, cut back the clump to half in spring.
- Harvest only the stem while leaving the roots intact and clear any dead leaves.
- Give plenty of compost and free draining soil to ensure it lasts year after year.

SHAPE			
TREE	PALM	SHRUB	CLUMP
VINE	HERB	**GRASS**	GROUND COVER

GROWTH (METRES)					
H	1	W	.5	SPACE	3

PRODUCTIVE SEASON											
J	**F**	**M**	**A**	**M**	**J**	**J**	**A**	**S**	**O**	**N**	**D**

LIGHT REQUIREMENTS		
SHADE	**PART SHADE**	FULL SUN

LEMON MYRTLE, *Backhousia citriodora*

- Native to Australia
- Dense compact tree and great tall hedge
- Superb intensely citrus fragrance of its flowers and leaves
- Source of calcium, high in folate, Vitamin A and Vitamin E
- Leaves add a nice lemon flower to tea and flavour to food.

SHAPE			
TREE	PALM	**SHRUB**	CLUMP
VINE	HERB	GRASS	GROUND COVER

GROWTH (METRES)					
H	4	W	2.5	SPACE	3

PRODUCTIVE SEASON											
J	**F**	**M**	**A**	**M**	**J**	**J**	**A**	**S**	**O**	**N**	**D**

LIGHT REQUIREMENTS		
SHADE	PART SHADE	**FULL SUN**

MADAGASCAR BEAN, *Phaseolus lunatus*

- Perennial (short lived) vine that produces a large bean that can be eaten fresh or allowed to mature and used in soups or stews.
- Beans get quite large and when cooked taste like potato.
- Most abundant after the second year of growth.
- Produces for 9 months of the year.

SHAPE			
TREE	PALM	SHRUB	CLUMP
VINE	HERB	GRASS	GROUND COVER

GROWTH (METRES)					
H	3	W	6	SPACE	2

PRODUCTIVE SEASON											
J	**F**	**M**	**A**	**M**	**J**	**J**	**A**	**S**	**O**	N	D

LIGHT REQUIREMENTS		
SHADE	**PART SHADE**	FULL SUN

MEXICAN TARRAGON, *Tagetes lucida*

- Small herb with aniseed flavour
- Used in flavouring salads, smoothies & desserts
- Hardy perennial
- Propagate by division

SHAPE			
TREE	PALM	SHRUB	CLUMP
VINE	**HERB**	GRASS	GROUND COVER

GROWTH (METRES)					
H	.5	W	.2	SPACE	1

PRODUCTIVE SEASON											
J	**F**	**M**	**A**	**M**	**J**	**J**	**A**	**S**	**O**	**N**	**D**

LIGHT REQUIREMENTS		
SHADE	**PART SHADE**	**FULL SUN**

MINT, *Mentha sp.*

- Doesn't like the heat too much so may need to be shaded through the wet season.
- Likes moisture but not 'wet feet'.
- Propagated from cutting or root separation.
- Many types; chocolate mint, spearmint, pennyroyal.

SHAPE			
TREE	PALM	SHRUB	CLUMP
VINE	HERB	GRASS	**GROUND COVER**

GROWTH (METRES)					
H	.1	W	-	SPACE	.5

PRODUCTIVE SEASON											
J	**F**	**M**	**A**	**M**	**J**	**J**	**A**	**S**	**O**	N	D

LIGHT REQUIREMENTS		
SHADE	**PART SHADE**	FULL SUN

MORINGA, *Moringa oleifera*

- Fast growing legume tree.
- Edible flowers, young beans, roots and leaves.
- Highly medicinal food and functional plant.
- Used as chop 'n' drop mulch.
- Can be grown as a living fence or trellis.
- Highly nutritious, packed with vitamins, minerals, and antioxidants.

SHAPE			
TREE	PALM	SHRUB	CLUMP
VINE	HERB	GRASS	GROUND COVER

GROWTH (METRES)					
H	4.5	W	3	SPACE	3

PRODUCTIVE SEASON											
J	F	M	A	M	J	J	**A**	**S**	**O**	**N**	**D**

LIGHT REQUIREMENTS		
SHADE	**PART SHADE**	**FULL SUN**

MOTHER OF ALL HERB, *Coleus amboinicus*

- Perennial ground cover.
- Thick furry leaf.
- Used as a great oregano substitute.
- Good used as a tea and restless nights.
- Anti-viral properties.
- Drought tolerant.

SHAPE			
TREE	PALM	SHRUB	CLUMP
VINE	**HERB**	GRASS	**GROUND COVER**

GROWTH (METRES)					
H	.5	W	1	SPACE	1

PRODUCTIVE SEASON											
J	**F**	**M**	**A**	**M**	**J**	**J**	**A**	**S**	**O**	**N**	**D**

LIGHT REQUIREMENTS		
SHADE	PART SHADE	FULL SUN

MUCUNA, *Mucuna pruriens*

- Annual vine that grows bean pods that dry with a hard shell.
- Edible bean needs to be soak and peeled before eating.
- Has a dopamine precursor that makes you happy.
- Hair around unripe bean can irritate.

SHAPE			
TREE	PALM	SHRUB	CLUMP
VINE	HERB	GRASS	GROUND COVER

GROWTH (METRES)					
H	1	W	1	SPACE	1

PRODUCTIVE SEASON											
J	**F**	**M**	A	M	J	J	A	S	O	**N**	**D**

LIGHT REQUIREMENTS		
SHADE	**PART SHADE**	FULL SUN

MUGWORT, *Artemisia vulgaris*

- Great as anti-parasitic around chickens.
- It grows as a hardy, clumping perennial with tall, leafy stems.
- Traditionally used for digestion, relaxation, and menstrual support.
- Good for dreaming and helps sleeping.
- Grows large so needs some space.

SHAPE			
TREE	PALM	**SHRUB**	CLUMP
VINE	**HERB**	GRASS	GROUND COVER

GROWTH (METRES)					
H	1	W	2	SPACE	1

PRODUCTIVE SEASON											
J	**F**	**M**	**A**	**M**	**J**	**J**	**A**	**S**	**O**	**N**	**D**

LIGHT REQUIREMENTS		
SHADE	**PART SHADE**	FULL SUN

NEEM, *Azadirachta indica*

- Known for its antibacterial, anti fungal, and anti-inflammatory properties.
- The leaves, seeds, and bark are used to treat skin conditions, boost immunity, and improve oral health.
- It is a fast-growing evergreen tree with small, fragrant white flowers and serrated leaves.
- Can become weedy if not managed.

SHAPE			
TREE	PALM	SHRUB	CLUMP
VINE	HERB	GRASS	GROUND COVER

GROWTH (METRES)					
H	4	W	4	SPACE	3

PRODUCTIVE SEASON											
J	**F**	**M**	**A**	**M**	**J**	**J**	**A**	**S**	**O**	**N**	**D**

LIGHT REQUIREMENTS		
SHADE	**PART SHADE**	FULL SUN

NEW GUINEA BEAN, *Lagenaria siceraria*

- Annual vine with long zucchini type bean
- Very tasty & versatile vegetable used in cooking or cakes.
- Very vigorous.
- Beans can grow up to 1m long.
- Mature gourds can be used for craft such as water bottles and pipes.

SHAPE			
TREE	PALM	SHRUB	CLUMP
VINE	HERB	GRASS	GROUND COVER

GROWTH (METRES)					
H	2	W	4	SPACE	1

PRODUCTIVE SEASON											
J	**F**	**M**	**A**	M	J	J	A	S	**O**	**N**	**D**

LIGHT REQUIREMENTS		
SHADE	**PART SHADE**	**FULL SUN**

OKINAWA SPINACH, *Gynura crepioides*

- Perennial ground cover / shrub.
- Produces spinach all year round, which can be eaten raw or cooked.
- Has an ornamental benefit with its green top and purple underside.
- Grown from cuttings 10-25cm long.

SHAPE			
TREE	PALM	SHRUB	CLUMP
VINE	HERB	GRASS	**GROUND COVER**

GROWTH (METRES)					
H	.5	W	1.5	SPACE	2.5

PRODUCTIVE SEASON											
J	**F**	**M**	**A**	**M**	**J**	**J**	**A**	**S**	**O**	**N**	**D**

LIGHT REQUIREMENTS		
SHADE	**PART SHADE**	FULL SUN

PARSLEY, *Petroselinum crispum*

- Versatile herb.
- Needs good drainage, best to keep in a pot to last the tropical wet season.
- It is packed with vitamins A, C, and K, supporting immunity, bone health.
- It comes in two main varieties: curly-leaf and flat-leaf.
- Traditionally, it is used for digestive health and as a diuretic.

SHAPE			
TREE	PALM	SHRUB	CLUMP
VINE	**HERB**	GRASS	GROUND COVER

GROWTH (METRES)					
H	.5	W	.5	SPACE	.2

PRODUCTIVE SEASON											
J	**F**	**M**	**A**	**M**	**J**	**J**	**A**	**S**	**O**	**N**	**D**

LIGHT REQUIREMENTS		
SHADE	**PART SHADE**	FULL SUN

PLANTAIN, *Plantago sp.*

- Medicinal herb as wound healer
- Many varieties that grow in different climates
- Contains allantoin, helps new skin cells grow
- Relieves bites, stings and mild pain
- Struggles in the humid tropics, save in a pot over the wet season

SHAPE			
TREE	PALM	SHRUB	CLUMP
VINE	**HERB**	GRASS	GROUND COVER

GROWTH (METRES)					
H	1	W	1	SPACE	.5

PRODUCTIVE SEASON											
J	F	**M**	**A**	**M**	**J**	**J**	**A**	**S**	**O**	**N**	**D**

LIGHT REQUIREMENTS		
SHADE	**PART SHADE**	FULL SUN

PEANUT, *Arachis sp.*

- The edible peanut (A. hypogaea), widely known and consumed, grows as an annual plant in tropical regions.
- The Pinto Peanut (A. pintoi), a perennial ground cover, a beneficial nitrogen fixer for orchards but is generally not eaten
- Both species have yellow, pea-like flowers and are valuable plants to cultivate.

SHAPE			
TREE	PALM	SHRUB	CLUMP
VINE	HERB	GRASS	**GROUND COVER**

GROWTH (METRES)					
H	.5	W	.5	SPACE	.25

PRODUCTIVE SEASON											
J	F	**M**	**A**	**M**	**J**	**J**	**A**	**S**	**O**	**N**	**D**

LIGHT REQUIREMENTS		
SHADE	**PART SHADE**	FULL SUN

PIGEON PEA, *Cajanus cajan*

- Legume support tree.
- Abundant producer of a lentil type bean
- Bee fodder.
- Sprouted bean's are quite tasty.
- Chop 'n' drop mulch just before flower to benefit from the most nitrogen release.
- Grown in the food forest to support other trees and improve soil.

SHAPE			
TREE	PALM	**SHRUB**	CLUMP
VINE	HERB	GRASS	GROUND COVER

GROWTH (METRES)					
H	3	W	3	SPACE	1

PRODUCTIVE SEASON											
J	**F**	M	A	M	J	J	A	S	O	**N**	**D**

LIGHT REQUIREMENTS		
SHADE	PART SHADE	**FULL SUN**

QLD ARROWROOT, *Canna edulis*

- 'Chop and Drop' for mulch.
- The large round red rhizomes can be eaten cooked as you would a potato.
- Good marginal water plant.
- Partially dies back in winter.

SHAPE			
TREE	PALM	SHRUB	**CLUMP**
VINE	HERB	GRASS	GROUND COVER

GROWTH (METRES)					
H	2	W	1	SPACE	1

PRODUCTIVE SEASON											
J	**F**	**M**	**A**	M	J	J	**A**	**S**	**O**	**N**	**D**

LIGHT REQUIREMENTS		
SHADE	**PART SHADE**	FULL SUN

ROSEMARY, *Salvia rosmarinus*

- Aromatic culinary herb.
- Likes very good drainage.
- Drought tolerant.
- Propagation from cutting.
- Grows well in raised unglazed clay pots in the wet tropics or under an eve on western side well drained garden bed.

SHAPE			
TREE	PALM	SHRUB	**CLUMP**
VINE	HERB	GRASS	GROUND COVER

GROWTH (METRES)					
H	1	W	.75	SPACE	.75

PRODUCTIVE SEASON											
J	**F**	**M**	**A**	**M**	**J**	**J**	**A**	**S**	**O**	**N**	**D**

LIGHT REQUIREMENTS		
SHADE	**PART SHADE**	FULL SUN

RUNGI, *Rungia klossii*

- A perennial spinach that tastes like mushroom!
- Slow growing but hardy plant.
- Likes the shade.
- Best kept pruned to make shrub-like.
- Propagated from cuttings.
- Likes well drained soil.

SHAPE			
TREE	PALM	**SHRUB**	CLUMP
VINE	HERB	GRASS	GROUND COVER

GROWTH (METRES)					
H	1	W	.75	SPACE	.75

PRODUCTIVE SEASON											
J	**F**	**M**	**A**	**M**	**J**	**J**	**A**	**S**	**O**	**N**	**D**

LIGHT REQUIREMENTS		
SHADE	PART SHADE	FULL SUN

SAWTOOTH CORIANDER, *Eryngium foetidum*

- Perennial alternative to the annual coriander.
- Best used fresh as it looses its flavour when cooked.
- Likes a moist location.
- Can be seen popping up in lawns.
- Remove the seed head regularly so the plant pots energy into leaf growth.

SHAPE			
TREE	PALM	SHRUB	CLUMP
VINE	**HERB**	GRASS	GROUND COVER

GROWTH (METRES)					
H	.2	W	.8	SPACE	2.5

PRODUCTIVE SEASON											
J	**F**	**M**	**A**	**M**	**J**	**J**	**A**	**S**	**O**	**N**	**D**

LIGHT REQUIREMENTS		
SHADE	PART SHADE	FULL SUN

SAMBUNG, *Gynura procumbens*

- Perennial ground cover / shrub.
- Also named 'life extender' for its medicinal qualities.
- Produces spinach all year round.
- Known to lower cholesterol.
- Rich in protein, iron, potassium, calcium, vitamin A.
- Grown from cuttings 10-25cm long
- Can be eaten raw or cooked.

SHAPE			
TREE	PALM	SHRUB	CLUMP
VINE	HERB	GRASS	**GROUND COVER**

GROWTH (METRES)					
H	.5	W	1.5	SPACE	2.5

PRODUCTIVE SEASON											
J	**F**	**M**	**A**	**M**	**J**	**J**	**A**	**S**	**O**	**N**	**D**

LIGHT REQUIREMENTS		
SHADE	PART SHADE	FULL SUN

SUN JEWEL, *Portulaca oleracea*

- Delicate perennial ground cover with colour flower.
- Various coloured varieties which all are edible as a leafy green.
- Has a slight sour / salty taste.
- Rich source of omega-3.
- Drought tolerant.

SHAPE			
TREE	PALM	SHRUB	CLUMP
VINE	**HERB**	GRASS	**GROUND COVER**

GROWTH (METRES)					
H	.2	W	1	SPACE	.2

PRODUCTIVE SEASON											
J	**F**	**M**	**A**	**M**	**J**	**J**	**A**	**S**	**O**	**N**	**D**

LIGHT REQUIREMENTS		
SHADE	**PART SHADE**	FULL SUN

SWEET LEAF, *Sauropus androgynus*

- Perennial bush.
- Protein source.
- Fresh tips are like asparagus and leaves have a sweet nutty taste.
- Best planted as a hedge.
- Leaves are good in pesto.
- Requires regular pruning.
- Propagate from cutting or seed.

SHAPE			
TREE	PALM	**SHRUB**	CLUMP
VINE	**HERB**	GRASS	GROUND COVER

GROWTH (METRES)					
H	1.5	W	4	SPACE	.5

PRODUCTIVE SEASON											
J	**F**	**M**	**A**	**M**	**J**	**J**	**A**	**S**	**O**	**N**	**D**

LIGHT REQUIREMENTS		
SHADE	**PART SHADE**	FULL SUN

SWEET POTATO, *Ipomoea batatas*

- Perennial ground cover that can be treated like an annual crop of potato.
- Plant in October and harvest from May.
- Many different varieties and colours.
- Propagate new plants from tips to get best crop.
- Leaves can be eaten of some varieties.
- Likes well drained loose soil with lots of organic matter to prodoce large tubers.

SHAPE			
TREE	PALM	SHRUB	CLUMP
VINE	HERB	GRASS	**GROUND COVER**

GROWTH (METRES)					
H	.3	W	3	SPACE	.2

PRODUCTIVE SEASON											
J	F	M	A	**M**	**J**	**J**	**A**	**S**	O	N	D

LIGHT REQUIREMENTS		
SHADE	**PART SHADE**	FULL SUN

TEA PLANT, *Camellia sinensis*

- Leaves used to make a cuppa tea.
- Leaves can be dried or used fresh.
- Has a pretty white and yellow flower.
- Best kept as bush, hedging required twice a year.
- Performs best in moist, acidic, well drained soil.

SHAPE			
TREE	PALM	**SHRUB**	CLUMP
VINE	HERB	GRASS	GROUND COVER

GROWTH (METRES)					
H	1	W	1	SPACE	.5

PRODUCTIVE SEASON											
J	**F**	**M**	**A**	**M**	**J**	**J**	**A**	**S**	**O**	**N**	**D**

LIGHT REQUIREMENTS		
SHADE	**PART SHADE**	FULL SUN

THAI BASIL, *Ocimum basilicum*

- Large perennial basil with a aniseed flavour.
- Used in Southeast Asian cooking as a garnish.
- Best kept pruned as shrub.
- Good bee fodder.
- Propagated from seed or cutting.
- Pinch the flowers to improve flavour.
- Harvest periodically to encourage growth.

SHAPE			
TREE	PALM	**SHRUB**	CLUMP
VINE	**HERB**	GRASS	GROUND COVER

GROWTH (METRES)					
H	1	W	.75	SPACE	.75

PRODUCTIVE SEASON											
J	**F**	**M**	**A**	**M**	**J**	**J**	**A**	**S**	**O**	**N**	**D**

LIGHT REQUIREMENTS		
SHADE	**PART SHADE**	FULL SUN

TOOTHACHE PLANT, *Spilanthes acmella*

- It is known for its unique numbing effect, often used to relieve toothaches and oral discomfort.
- The plant produces vibrant yellow flowers with a distinctive spicy or tingling taste when consumed.
- Excellent for teething babies

SHAPE			
TREE	PALM	SHRUB	CLUMP
VINE	**HERB**	GRASS	GROUND COVER

GROWTH (METRES)					
H	.5	W	.5	SPACE	.2

PRODUCTIVE SEASON											
J	**F**	**M**	**A**	**M**	**J**	**J**	**A**	**S**	**O**	**N**	**D**

LIGHT REQUIREMENTS		
SHADE	**PART SHADE**	FULL SUN

TURMERIC, *Curcuma longa*

- Edible tubers, for fragrant kitchen spice
- Great anti- inflammatory.
- Abundant leaf growth, its a great cover crop and soil improver.
- Some varieties include, yellow, gold and black.
- Harvest rhizomes when the leaves have died back however you can 'bandicoot' pieces earlier.

SHAPE			
TREE	PALM	SHRUB	**CLUMP**
VINE	**HERB**	GRASS	GROUND COVER

GROWTH (METRES)					
H	1	W	1	SPACE	1

PRODUCTIVE SEASON											
J	F	M	A	M	J	**J**	**A**	**S**	**O**	**N**	D

LIGHT REQUIREMENTS		
SHADE	**PART SHADE**	FULL SUN

VANILLA, *Vanilla planifolia*

- Vine is actually an orchid
- Grows well in the shade
- Each flower needs to be hand pollinated which has only a 12 year opening period.
- Will then take 6-9 months to mature into a vanilla pod.
- Will not flower in low light conditions.
- Will start flowering after 3 years.

SHAPE			
TREE	PALM	SHRUB	CLUMP
VINE	HERB	GRASS	GROUND COVER

GROWTH (METRES)					
H	1.5	W	4	SPACE	.5

PRODUCTIVE SEASON											
J	F	M	A	M	J	J	A	S	**O**	**N**	**D**

LIGHT REQUIREMENTS		
SHADE	**PART SHADE**	FULL SUN

VAP CA, *Houttuynia cordata*

- Strong, fish-like aroma; used in Southeast Asian cuisine and medicine.
- Rich in antioxidants, boosts immunity, reduces inflammation, aids digestion.
- Heart-shaped leaves, small white flowers, grows as ground cover.
- Thrives in moist, shady areas; valued for culinary and medicinal uses.

SHAPE			
TREE	PALM	SHRUB	CLUMP
VINE	HERB	GRASS	**GROUND COVER**

GROWTH (METRES)					
H	.5	W	.5	SPACE	.2

PRODUCTIVE SEASON											
J	**F**	**M**	**A**	**M**	**J**	**J**	**A**	**S**	**O**	**N**	**D**

LIGHT REQUIREMENTS		
SHADE	PART SHADE	FULL SUN

VETIVER GRASS, *Chrysopogon zizanioides*

- Clumping rass that doesn't spread.
- Landscaping plant used as garden edging.
- Deep root system.
- Great chop and drop mulch source and supports trees to grow.
- Erosion control on steep slopes.
- Doesn't handle shade.
- Root oil used in perfume.

SHAPE			
TREE	PALM	SHRUB	CLUMP
VINE	HERB	**GRASS**	GROUND COVER

GROWTH (METRES)					
H	1	W	1	SPACE	3

PRODUCTIVE SEASON											
J	**F**	**M**	**A**	**M**	**J**	**J**	**A**	**S**	**O**	**N**	**D**

LIGHT REQUIREMENTS		
SHADE	**PART SHADE**	**FULL SUN**

VIETNAMESE MINT, *Persicaria odorata*

- Peppery, citrusy flavor with a hint of coriander.
- It is a fast-growing, spreading herb, thrives in warm, moist conditions.
- Traditionally used for digestion, inflammation, and fever relief due to its antibacterial and cooling properties.
- Natural pest deterrent and grows well with other moisture-loving plants.

SHAPE			
TREE	PALM	SHRUB	CLUMP
VINE	**HERB**	GRASS	GROUND COVER

GROWTH (METRES)					
H	.5	W	.5	SPACE	.2

PRODUCTIVE SEASON											
J	**F**	**M**	**A**	**M**	**J**	**J**	**A**	**S**	**O**	**N**	**D**

LIGHT REQUIREMENTS		
SHADE	PART SHADE	FULL SUN

WING BEAN, *Psophocarpus tetragonolobus*

- Semi-perennial vines that produces a green bean that can be eaten fresh.
- Has a pretty blue flower that is sweet and good in salads.
- Plants dies back in the dry season and return in the wet.
- Produces for 6-9 months of the year.

SHAPE			
TREE	PALM	SHRUB	CLUMP
VINE	HERB	GRASS	GROUND COVER

GROWTH (METRES)					
H	3	W	1	SPACE	2

PRODUCTIVE SEASON											
J	**F**	**M**	**A**	M	J	J	**A**	**S**	**O**	**N**	**D**

LIGHT REQUIREMENTS		
SHADE	**PART SHADE**	**FULL SUN**

WATER CHESTNUT, *Eleocharis dulcis*

- Crisp, sweet corms used in stir-fries, salads, and soups.
- It grows in shallow water, wet soil or shallow pond.
- Propagates by corm division and matures in about six months.
- Rich in fibre and antioxidants.
- Harvested when leaves turn yellow.

SHAPE			
TREE	PALM	SHRUB	**CLUMP**
VINE	HERB	GRASS	GROUND COVER

GROWTH (METRES)					
H	.5	W	.5	SPACE	.2

PRODUCTIVE SEASON											
J	**F**	**M**	**A**	**M**	**J**	**J**	**A**	**S**	**O**	**N**	**D**

LIGHT REQUIREMENTS		
SHADE	**PART SHADE**	FULL SUN

YACON, *Smallanthus sonchifolius*

- It is a tuberous root vegetable with a sweet, crisp texture, often eaten raw or used in syrups.
- Known for its high fructooligosaccharide content, it acts as a natural sweetener and supports gut health.
- Has bright yellow daisy-like flowers.
- Propagate the cormlets and not the main tuber.

SHAPE			
TREE	PALM	SHRUB	CLUMP
VINE	**HERB**	GRASS	GROUND COVER

GROWTH (METRES)					
H	1	W	.5	SPACE	.2

PRODUCTIVE SEASON											
J	F	M	A	**M**	**J**	**J**	**A**	S	O	N	D

LIGHT REQUIREMENTS		
SHADE	**PART SHADE**	FULL SUN

YLANG-YLANG, *Cananga odorata*

- Native to SE Asia, it is very popular for its extremely fragrant flowers and weeping branches.
- The essential oil qualities are as antiseptic, aphrodisiac and a tonic to the nervous system.
- Will grow tall, so best keep pruned to access flowers.
- Will begin flowering after 3-4 years.

SHAPE			
TREE	PALM	SHRUB	CLUMP
VINE	HERB	GRASS	GROUND COVER

GROWTH (METRES)					
H	8	W	4	SPACE	4

PRODUCTIVE SEASON											
J	F	M	**A**	**M**	**J**	J	A	S	O	N	D

LIGHT REQUIREMENTS		
SHADE	PART SHADE	**FULL SUN**

ZINNIA, *zinnia sp.*

- Semi-perennial edible flower up to 5cm.
- Various species of different colours.
- Adds lots of colour to the garden.
- Self propagates from seed or from cutting.

SHAPE			
TREE	PALM	SHRUB	CLUMP
VINE	**HERB**	GRASS	GROUND COVER

GROWTH (METRES)					
H	.5	W	.5	SPACE	.2

PRODUCTIVE SEASON											
J	**F**	**M**	**A**	**M**	**J**	**J**	**A**	**S**	**O**	**N**	**D**

LIGHT REQUIREMENTS		
SHADE	**PART SHADE**	FULL SUN

Bamboo list

Name	Height	Min. Temp	Description
Bambusa arnhemica	12m	0°C	Green culms, small leaves, strong near-solid stems. Used by Aboriginal people for didgeridoos.
Bambusa boniopsis	4m	-7°C	Dense foliage, slightly weeping tops, delicate look. Great for screening or small features.
Bambusa chungii 'Barbelletta'	8 to 10m	-4°C	Compact B. chungii with white powdery culms. Ideal for decoration or privacy screens.
Bambusa chungii (White Bamboo)	10 to 12m	-4°C	Ornamental bamboo with white bloom-covered culms. Perfect for privacy screens or features.
Bambusa dolichomerithalla	10m	-2°C	Light green culms with dark green striations, turning yellow with age. A beautiful feature plant.
Bambusa dolichomerithalla 'Blue' (Doli Blue)	8m	-5°C	Straight green culms with white powdery coating. Cold-tolerant, grows well in southern regions.
Bambusa dolichomerithalla 'Silverstripe'	8m	-2°C	Dark green culms with silver striations. Excellent ornamental choice.
Bambusa eutuldoides var. Viridi Vittata (China Gold)	6 to 8m	-9°C	Lemon-yellow culms with green striations. Great for screening or decoration.
Bambusa guangxiensis (Chinese Dwarf)	4 to 8m	-2°C	Bushy bamboo, ideal for hedging and privacy screens. Trim to 2.5–4m for best results.
Bambusa heterostachya	6m	-2°C	Weeping ornamental bamboo. Suitable for shade or as a feature plant.
Bambusa heterostachya 'Variegata' (Variegated Malay Dwarf)	3 to 4m	-2°C	Dense bamboo with cream-striped leaves. One of the best for hedging.
Bambusa lako (Timor Black)	10m	-2°C	Glossy black culms with green striations. A striking feature plant.
Bambusa maculata (Pring Tutul)	20m	-2°C	Produces mottled timber, used in Indonesia for furniture.
Bambusa malingensis	12m	-7°C	Dense clumps, upright culms, tolerates coastal winds.
Bambusa multiplex cv. Alphonse Karr	5m	-12°C	Yellow culms with green stripes. Good for windbreaks or privacy screens.
Bambusa multiplex cv. Goldstripe	4 to 5m	-12°C	Small bamboo with gold-striped culms. Perfect for screening in small gardens.
Bambusa nana - Thai Beauty (Large Leaf)	8 to 10m	-2°C	Tight clumping, smooth culms, lush weeping tops. A stunning bamboo.
Bambusa nana - Thai Silk (Fine Leaf)	6 to 7m	-2°C	Tiny leaves, tight clumping, near-solid culms. A must-have for garden enthusiasts.
Bambusa nutans	10m	-5°C	Straight, sturdy culms with edible shoots. Tight clumping habit.
Bambusa oldhamii	15 to 18m	-4°C	Hardy, fast-growing bamboo. Ideal for windbreaks, privacy screens, and edible shoots.
Bambusa oliveriana	6 to 8m	-5°C	Near-solid culms, great for screening, features, or timber. Tight clumping habit.
Bambusa textilis (Weaver's Bamboo)	10 to 12m	-12°C	Straight culms, branch-free lower half, long internodes. A beautiful ornamental.
Bambusa textilis var. Fusca	10 to 12m	-12°C	Similar to Gracilis but larger. Smooth green culms, tidy upright growth.
Bambusa textilis var. Glabra	10 to 12m	-12°C	Thin-walled ornamental bamboo with white lower internodes.
Bambusa textilis var. Gracilis (Slender Weaver Bamboo)	6 - 8m	-12°C	Tidy upright bamboo. Great for privacy screens or feature plantings. Popular in Australia.
Bambusa ventricosa (Buddha's Belly)	6-10m	-9°C	Develops swollen internodes in harsh conditions. A unique ornamental.
Bambusa vulgaris cv. Vittata (Painted Bamboo)	15m	-2°C	Lemon-yellow culms with green stripes, open clumping, weeping stems.
Bambusa vulgaris cv. Wamin (Giant Buddha's Belly)	5 to 6m	-2°C	Large swollen internodes, weeping culms. An attractive ornamental.
Bambusa vulgaris cv. Wamin striated (Striated Buddha's Belly)	4 to 5m	-2°C	Smaller 'Wamin' with light green culms and dark green striations. Great for decoration.
Cephalostachyum pergracile	6m	-2°C	Erect light green culms, foliage to the ground. Tight clumping, great for screening or features.
Dendrocalamopsis variostriata	4m	-2°C	Dense bamboo with fern-like leaves. Ideal for privacy screens or small features.
Dendrocalamus asper (Indonesian Clone)	30m	-5°C	Enormous straight culms with velvet-like hair. Grown for timber and edible shoots. Best for acreage.
Dendrocalamus asper cv. 'Hitam' (Black Asper or Betung Hitam)	30m	-5°C	Black-culmed giant, fast-growing. Produces excellent timber and edible shoots.
Dendrocalamus brandisii	36m	-4°C	Tallest bamboo in the world. Thick-walled culms, great for timber and edible shoots.
Dendrocalamus brandisii var. 'Black' (Black Brandisii)	36m	-5°C	Green culms turn black with age. Vigorous grower, produces quality timber and edible shoots.
Dendrocalamus giganteus (Giant Bamboo)	30m	-2°C	One of the largest bamboos. Needs tropical conditions for full potential. Produces premium timber.
Dendrocalamus latiflorus	25m	-4°C	Light green culms, large leaves, sweet edible shoots.
Dendrocalamus latiflorus cv. 'Mei Nung'	24m	-4°C	Light green culms with dark green stripes, large leaves. A stunning ornamental with edible shoots.
Dendrocalamus maroochy	15m	-4°C	Green culms with yellow striations, large leaves, slightly arching. Good for timber and edible shoots.
Dendrocalamus membranaceus	15m	-4°C	Large bamboo, small leaves, bushy growth. Erect culms with weeping tops.
Dendrocalamus minor	8m	-5°C	Glossy large leaves, smooth culms. A beautiful ornamental.
Dendrocalamus minor var. Amoenus (Ghost Bamboo)	8m	-5°C	Lime-green culms with white powder, dark green striations. Tidy, upright growth with arching tops.
Dendrocalamus sikkimensis	15m	-2°C	Rusty orange culms, large leaves. Produces great timber and edible shoots.
Gigantochloa apus (Tabashir bamboo)	20m	-2°C	Culms contain Tabashir, used in Chinese medicine. Produces high-quality timber.
Gigantochloa atroviolacea (Java Black)	10 to 15m	-2°C	Sooty black culms, lush large leaves. A spectacular feature plant with great timber.
Gigantochloa atroviolacea var. 'Watupawan'	16m	-2°C	Straight green culms with black striations. Excellent timber bamboo for furniture.
Gigantochloa hasskarliana	6m	-2°C	Dense bamboo, large leaves, slightly weeping culms. Ideal for tropical gardens.
Gigantochloa luteostriata	5m	-2°C	Dark green culms with cream-striped leaves. Tight clumping, weeping tops.
Gigantochloa pseudoarundinacea var. 'Gombong Batu'	30m	-2°C	Vigorous grower, straight cream-striped culms. Produces first-grade timber and edible shoots.
Gigantochloa robusta	30m	-2°C	Giant bamboo, straight cream-striped culms, large leaves. Produces high-quality timber and shoots.
Gigantochloa sp. 'Awi Temen'	18m	-2°C	Dark green culms with cream stripes. Produces good-quality timber.
Gigantochloa sp. 'Hitam Hijau'	5m	-2°C	Weeping culms, large leaves, lush foliage. Perfect for tropical gardens.
Gigantochloa sp. 'Rachel Carson'	8 to 10m	-2°C	Straight, large-leafed bamboo with cream-striped culms. Produces edible shoots and quality timber.
Gigantochloa sp. G57	8m	-2°C	Medium-sized bamboo, great for screening and timber. Fast-growing with lush foliage.
Gigantochloa wrayi	6m	-2°C	Tight clumping, upright growth, thick foliage. Ideal for privacy and windbreaks.
Guadua 'Peruana'	12 to 15m	-2°C	Dark green culms with wide white nodes. Excellent feature plant with high-quality timber.
Guadua angustifolia 'Bicolour'	25m	-2°C	Culms with yellow striations. Produces exceptional-quality timber.
Guadua angustifolia (Colombian Giant Thorny)	25m	-2°C	Magnificent bamboo with exceptional timber. Thorny lower branches, open clumping.
Himalayacalamus porcatus (Nepalese Blue)	3m	-10°C	Delicate bamboo with fine leaves. Prefers cool temperate climates.
Nastus elatus (New Guinea Green)	15m	-2°C	Light green culms, soft fine leaves. Produces excellent edible shoots.
Neololeba atra (Iron Range Bamboo)	5m	0°C	Tight clumping, very large leaves. Prefers shade or warm, protected areas.
Schizostachyum brachycladum 'Yellow' (Sacred Bali Bamboo)	12m	-2°C	Bright yellow culms with green stripes, large yellow-striped leaves. Tropical appearance.
Schizostachyum dumetorum	5m	-2°C	Long pendulous culms. Great for draping over riverbanks or rock ledges.
Schizostachyum hainanense	8m	-2°C	Rough green culms, slender weeping tops. A fantastic ornamental.
Schizostachyum iraten (Bali Blowpipe Bamboo)	5m	-2°C	Green culms, bushy growth, large leaves. Ideal for small features or privacy screens.
Schizostachyum jaculans	6m	-2°C	Tight clumping bamboo with long internodes.
Schizostachyum lumampao	15m	-2°C	Large ornamental bamboo, very large leaves. Thin-walled timber used for matting.
Schizostachyum sp. 'Murray Island'	5m	-2°C	Native Australian bamboo, straight upright culms. Great for screening in small spaces.
Schizostachyum zollingeri	5m	-2°C	Tight clumping, straight culms, nodding tops. Produces useful timber and edible shoots.
Thyrsostachys siamensis - Clone I (Monastery Bamboo)	6m	-4°C	Tight clumping, weeping culms, fine leaves. Produces edible shoots and quality timber.
Thyrsostachys siamensis - Clone II (Monastery Bamboo)	13m	-4°C	Tight clumping, fine leaves, erect growth. Produces edible shoots and good timber.

Food forest

What Is a Food Forest?

A food forest is a diverse and resilient garden system where fruit trees, perennial vegetables, herbs, spices, root crops, and beneficial plants are intentionally grown together in layers—much like a natural forest. Designed using permaculture principles, a food forest mimics the structure and function of an ecosystem, aiming to regenerate soil, conserve water, support biodiversity, and produce food with minimal ongoing input. These systems include multiple layers: tall trees (canopy), smaller trees and shrubs (sub-canopy), herbaceous plants, ground covers, vines, and root crops. Each plant plays a distinct role in the ecosystem—some fix nitrogen, others attract pollinators, repel pests, provide mulch, or help build soil. Over time, a well-designed food forest becomes increasingly self-sustaining, requiring less maintenance and offering more abundance year after year.

The Role of Forest Guilds

At the heart of a successful food forest is the forest guild—a group of plants, animals, fungi, and microorganisms that interact in supportive ways. Each member has a specific role, and together they form a cooperative network that promotes resilience and balance. For example, deep-rooted plants mine minerals, ground covers shade the soil, flowering herbs attract pollinators, and fungi connect everything through underground mycelium.

Diversity is key. A rich mix of species allows the system to resist disease, drought, and pests. Over time, the food forest becomes not just a garden, but a living, breathing ecosystem that regenerates itself.

The 8 Layers of a Tropical Food Forest

This layered approach mimics the structure of a natural forest—stacking productivity vertically and below ground:

1. CANOPY (LARGE TREES AND PALMS)
2. MIDDLE STORY (MEDIUM UNDERSTORY TREES AND PALMS)
3. SHRUBS (LARGE HERBS, GINGERS, SMALL FRUITS)
4. HERBS (LEAFY GREENS, GRASSES, SMALL FRUITS & SPICES)
5. VINES (FRUITS, VEGETABLES, TOOLS & SPICES)
6. GROUND COVERS (HERBS, FRUITS & VEGETABLES)
7. RHIZOSPHERE (VEGETABLES, NUTS & SPICES)
8. MICROBIOLOGY (SUPPORT SYSTEM)

Sector Analysis and Zone plan

Planning, Design & Placement - Work from Patterns to Details
Before breaking ground, observe your site and begin designing with both function and plant needs in mind:

Zoning Plan

In permaculture design, Zoning is a method used to organize space based on how often you use different areas and how much attention they require. It helps create efficient systems that save time and energy. Zones are numbered from Zone 1 (closest to the house) - typically for herbs, veggies, and daily tasks - to Zone 5 (untouched wilderness) reserved for observation and natural regeneration.
Food forests usually belong in Zone 2 or 3 - close enough for regular visits, pruning, harvesting, and observation.

Sector Analysis

Sector analysis is the process of mapping the natural forces and influences that interact with your site. These include soil types, wind direction, sun paths, slope, shade, frost pockets, wildlife movement, and water flow. Understanding these sectors helps you make informed decisions about where to locate elements in your design—for example, placing windbreaks to protect sensitive plants, positioning taller trees to block harsh afternoon sun, or grouping water-loving plants in natural drainage areas. Sector analysis ensures your food forest works with nature, not against it and that you can devise an effective management strategy.

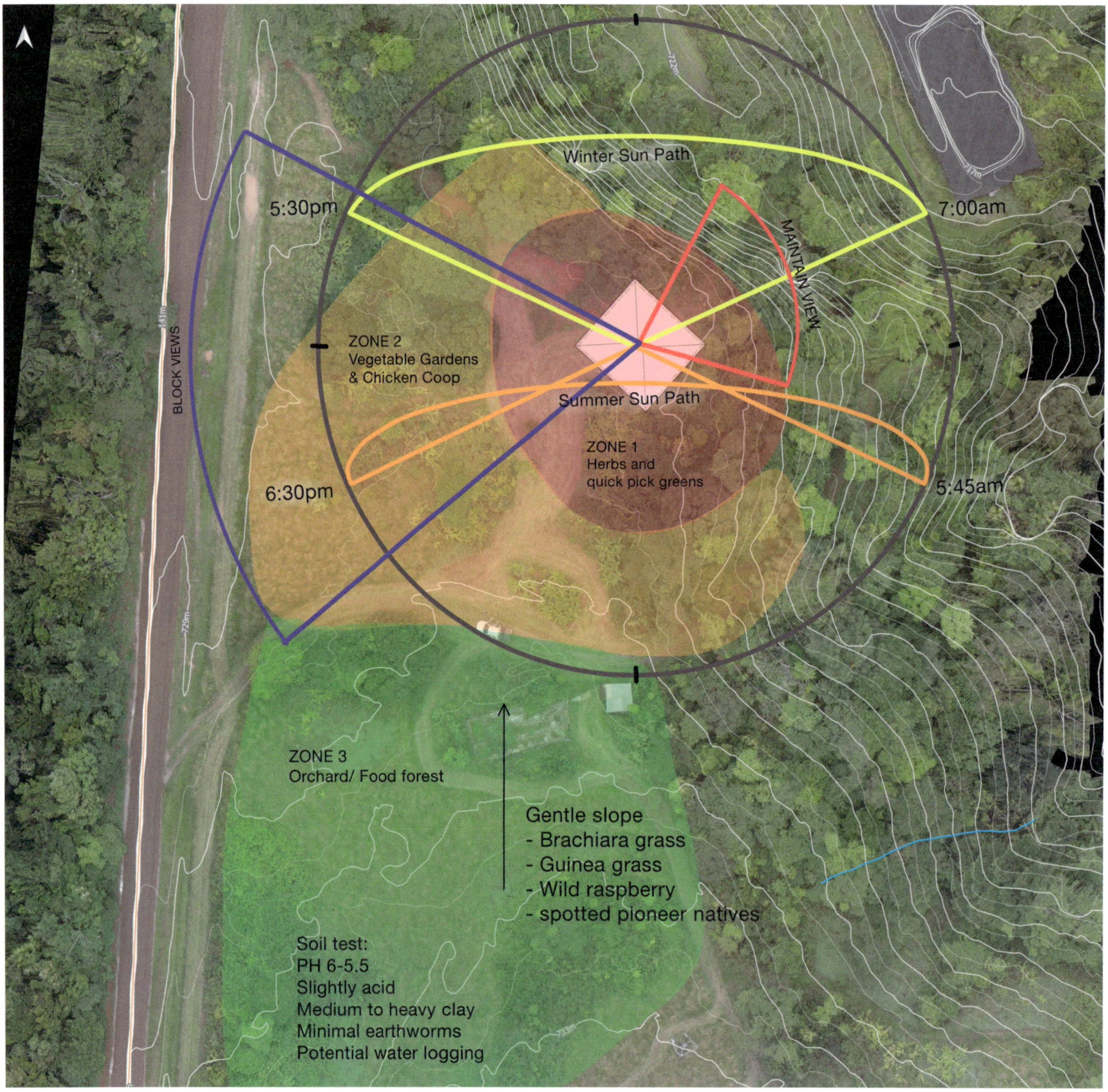

Understanding Soil Types: Texture, Structure, and Nutrient Balance

Soil is the foundation of every healthy garden, and understanding its texture and composition helps guide decisions around watering, fertilising, and improving plant performance.

Soil Texture: Sand, Silt, and Clay

Soil texture refers to the proportion of sand, silt, and clay particles that make up the mineral portion of the soil. These particle sizes determine how the soil behaves, and most soils are a combination of all three. The soil texture triangle is a visual tool used to classify soils (like sandy loam, silty clay, or loam) based on these proportions.

- **Sandy soils** are light and well-aerated, but they hold little water or nutrients. They can become hydrophobic (repelling water) when very dry and require regular additions of compost or organic matter to improve structure and moisture retention.
- **Clay soils** are heavier, with fine particles that hold water and nutrients well - but they can become dense, poorly drained, and prone to compaction.
- **Silt soils** are in between - smooth in texture and better at holding water than sand, but not as nutrient-dense as clay.

The term "light soil" typically refers to sandy soils that are easy to work, while "heavy soil" refers to clay-rich soil that can be more difficult to manage.

Why Texture Matters

Soil texture directly affects:

- **Water retention:** Clay holds more water; sand drains quickly.
- **Drainage and aeration:** Sandy soils are airy; clay can become waterlogged.
- **Workability:** Sandy soils are easy to dig; clay can be sticky or hard when dry.
- **Fertility:** Clay holds nutrients well; sand often leaches them.

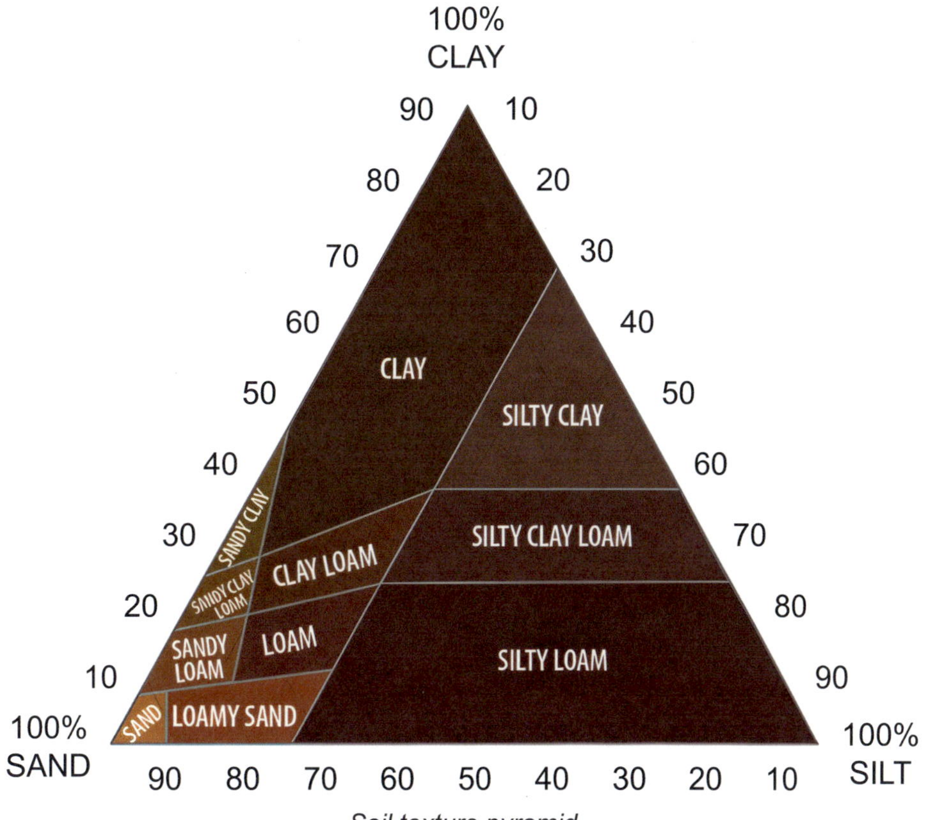

Soil texture pyramid

Testing Soil Texture at Home

You can assess your soil texture using the ribbon test:

1. Moisten a small handful of soil.
2. Roll it into a ball, then squeeze it between your fingers to form a ribbon.
3. The longer the ribbon before it breaks, the more clay is present.

Texture	Ribbon Length (mm)	Key Traits
Sandy	<15mm	Free-draining, low fertility, little structure.
Sandy Loam	15 - 25mm	Easy to work, moderate water/ nutrient retention.
Loam/Silty Loam	25mm	Balanced, fertile, ideal for most crops.
Clay Loam	40 - 50mm	Moisture-retentive, moderately heavy.
Clay	50 - 75mm	Holds water/nutrients well, can compact easily.
Heavy Clay	> 75mm	Dense, slow-draining, may restrict roots.

The Importance of Soil pH Testing

Soil pH affects how well plants can access nutrients. Most garden plants prefer a pH between 6.0 and 7.0. You can easily test your pH using a soil pH kit or meter.

Improving pH Based on Soil Type:

- **Sandy soils** are often acidic and low in magnesium—use dolomite lime, which adds both calcium and magnesium.
- **Clay soils** often already contain enough magnesium—use agricultural lime (calcium carbonate) instead of dolomite to improve structure without overloading magnesium.

PH testing on clay loam soils

Soil Management in Tropical Clay Soils

Most soils I work with are a mostly clay to silty clay and located in high-rainfall areas of Australia. These soils are often mineral-depleted, low in organic matter, and require consistent effort to improve. Building soil in these environments involves balancing minerals, increasing biology, and protecting the surface from tropical extremes.

Below is a general recommendation tailored to these conditions. However, it's important to do your own basic soil test first - this table may not suit all soils.

Input	Recommended Rate	Notes for Tropical High Rainfall
Lime (garden)	125–200 g/m² (1.25–2 t/ha) every 2–3 years	Stay on the high end of rate, check pH annually - rain will leach Ca
Soft Rock Phosphate	100–150 g/m² (1–1.5 t/ha) every 1–2 years	Great choice - doesn't leach, suits acidic clay soils
Gypsum (if used)	50 g/m² (0.5 t/ha) every 2–3 years	Only needed if avoiding lime or adding Ca without shifting pH
Gold Pellets / Organic NPK	125–150 g/m² every 6–8 weeks	Increase slightly - leaching of K and N is common
Blood & Bone	150–200 g/m² every 6–8 weeks	Boosts N levels; breaks down fast in humid conditions
Borax (Boron)	1–1.5 g/m² every 2 years	Leaches quickly - ight annual top-up may work better
Compost	2–5 cm topdress 2x/year	Breaks down rapidly—more frequent top-up beneficial
Biochar (charged)	0.5–1 kg/m² once	Excellent for water/nutrient retention in leaching soils
Mulch	5–10 cm deep, every 3–4 months	Vital for fungal activity and water regulation
Worm castings/vermicompost	0.5–1 kg per plant or 5–10% mix	Best at planting and pre-fruit set
Compost tea / worm tea	Every 2–3 weeks	Rebuilds microbial life post-heavy rainfall
Tri-Kelp (foliar spray)	Every 1–2 weeks	Supports plant stress recovery and trace uptake
Fish hydrolysate	Monthly	Highly beneficial in nutrient-leached soils
Mycorrhizal inoculant	One-time at planting	Ensure good root contact great for wet/dry cycles

When in Doubt – Get a Soil Mineral Test

If you're unsure about your soil's nutrient levels, mineral balance, or pH - a professional soil test is the most accurate way to assess it. It can reveal deficiencies or excesses in nutrients like phosphorus, potassium, calcium, magnesium, and trace elements - allowing you to fine-tune your inputs for optimal plant health and soil function.

Food Forest Plan

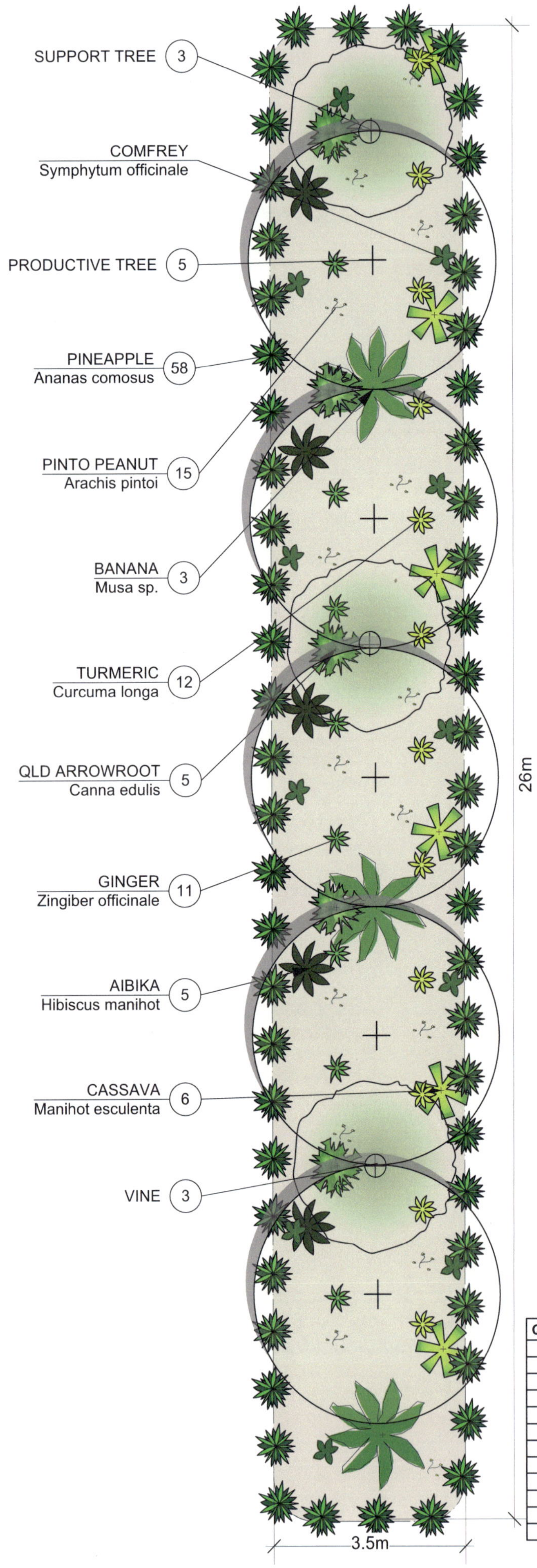

Create your planting list

Start by creating a list of all the plants you'd like to include in your garden. Choose species suited to your climate, personal tastes, and culinary interests. Use the plant spacing guidelines provided in this book to plan your layout. Measure the bed length and divide this length with spacing required for each plant. This will give to a total amount of plants needed.

Draw a planting plan

Begin by drawing your main species such as larger fruit and nut trees onto your beds. Then, fill the remaining spaces with support plants, perennial vegetables, herbs, spices, and root crops.

When planting vines, be strategic: they can overwhelm young trees if not managed. Instead, position vigorous climbers at the base of hardy or fast-growing support species and robust trees such as ice cream bean, jackfruit, or elderberry.

Choose a mix of support trees such as Eucalyptus (Eucalyptus grandis), Moringa (Moringa oleifera), Pigeon Pea (Cajanus cajan), Ice Cream Bean (Inga edulis), and Gliricidia (Gliricidia sepium) play key roles in tropical food forests by providing shade, mulch, nitrogen fixation, and structural support. These fast-growing species help improve soil, protect young plants, and accelerate ecosystem development.

Use small clumping plants to define the edges of your garden beds. These species create a clear visual and functional boundary while suppressing invasive grasses and reducing the need for ongoing maintenance. Plants like vetiver grass, lemongrass, and comfrey offer dense growth that blocks weeds and can be regularly cut back as chop-and-drop mulch to enrich the soil. Pineapple is another excellent edge plant—though not suitable for mulching, it contributes a productive fruit crop and forms a low barrier. Be sure to choose the non-spiky leaf variety of pineapple for ease of maintenance and safer handling.

The example plan on the left gives you an idea of a typical bed filled with plants and the amounts of plants required to fill a 26 x 3.5m area.

Qty	Common Name	Latin Name	Type	
5	AIBIKA	Hibiscus manihot	Shrub	Perennial spinach
3	BANANA	Musa sp.	Fruit	Good support plant
6	CASSAVA	Manihot esculenta	Shrub	Root crop and good leaf mulch
11	COMFREY	Symphytum officinale	Herb	Medicinal and soil builder
11	GINGER	Zingiber officinale	Herb	Valuable root crop
58	PINEAPPLE	Ananas comosus	Clump	Good edging plant
15	PINTO PEANUT	Arachis pintoi	Ground cover	Nitrogen fixing
5	PRODUCTIVE TREE	-	Tree	Choose your favorites
5	QLD ARROWROOT	Canna edulis	Clump	Root crop and biomass mulch
3	SUPPORT TREE	-	Trees	Choose a variety
12	TURMERIC	Curcuma longa	ROOT	Root crop and good soil builder
3	VINE	-	Vine	Choose your favorites

Begin building your plant stock by propagating the species that will support your system.

Propagation Techniques:

- **Cuttings:** Use 2–3 node cuttings from healthy plants. Strip lower leaves to reduce moisture loss. Place these cuttings in a bucket of water in preparation.
- **Root Division:** Carefully divide clumping perennials, ensuring each section has viable roots and shoots. Keep moist in damp newspaper or hessian sacks.
- **Seed Trays:** Sow seeds in a quality seed-raising mix. Label trays clearly to track varieties. These can be divided out at planting time.

Marking, Mowing & Earthworks

1. **Mark Beds & Paths**: Use stakes, string, or chalk spray paint to lay out planting zones.
2. **Mow Grass:** Cut grass low in both bed and path areas to reduce its strength before sheet mulching or tilling.
3. **Excavate Paths (optional):** On clay soils, improve drainage by mounding your garden beds. Remove the topsoil from pathways or the spaces between rows and mound it onto the beds. This technique is especially important for sensitive species like avocado, which require well-drained soil to thrive.
4. **Deep Rip:** Use a ripper or garden fork to loosen compacted soil through the centre of beds. This improves drainage and root penetration, especially in clay soils.
5. **Spread Minerals & Compost:** Apply minerals, fertilisers and compost.
6. **Sheet Mulch:** If the site is heavily grassed, lay cardboard over the beds as a weed barrier. Alternatively, multiple rounds of tilling can also help suppress weeds.

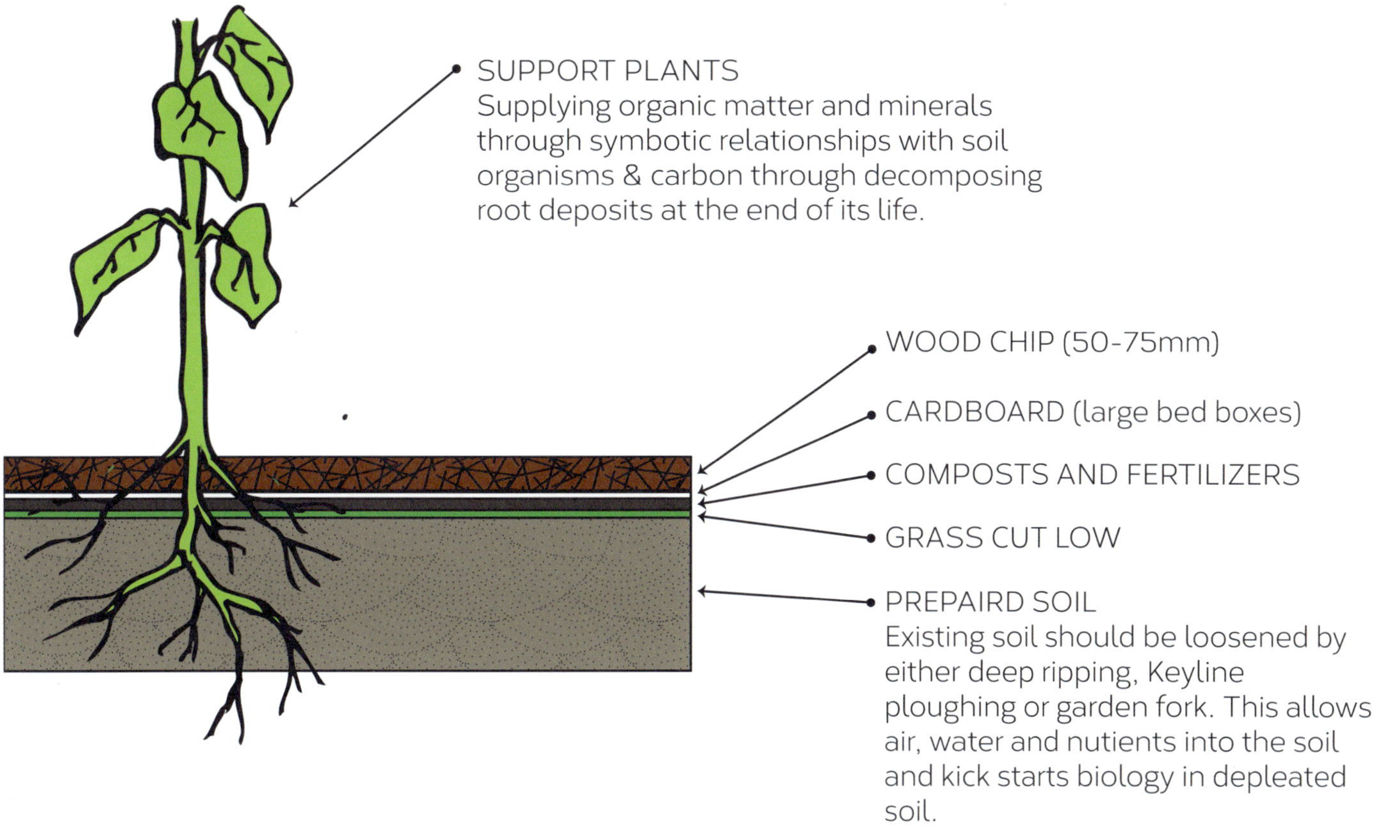

Bed preparation method

Define and Prepare Planting Locations

1. **Mark Plant Positions:** Use your layout to mark exact locations for trees based on your plan. Arranging the pots on top of the beds helps gauge positions and spacing of plants before they are planted.

2. **Dig Holes:**

- Use a 150 mm auger or post hole shovel to dig each hole to 1.5x the pot depth (about 30–40 cm)
- Break up clods to prevent air pockets.
- Avoid digging too far ahead - dig holes on the day of planting or the afternoon before.
- If using an auger (on clay), roughen the smooth hole walls with a crowbar to prevent root spiralling against the hard smooth hole.

Planting

- **Tri-Kelp Soak:** Soak cuttings, root divisions, and seedlings in a diluted tri-kelp solution for 1–2 hours before planting to stimulate root development.
- **Add Amendments:** Mix more compost, minerals and fertilisers directly into each planting hole.
- **Plant Trees:** Position trees so the root crown sits level with the soil surface. Backfill and firm the soil gently.
- **Plant Support Species:** Follow spacing guidelines. Push cuttings halfway into the soil; carefully place root divisions and seedlings into pre-dug holes.
- **Fertilise & Water In:** Apply an organic fertiliser around the base (not touching stems), then water thoroughly with a tri-kelp solution.

Mulching & Watering

- **Mulch Generously:** Apply a thick layer of wood chip mulch around trees, along beds. This suppresses weeds, retains moisture, and boosts fungal activity in the soil.
- **Irrigation:** Water well at planting. Maintain a regular watering schedule until plants are fully established. This will depend on your climate and soil type.

Succession Planning

In the early stages of your garden, Pigeon Pea (Cajanus cajan) acts as a pioneer species—offering essential functions such as nitrogen fixation, wind and sun protection, and a steady supply of leaf mulch. As a bonus, it also produces an abundance of edible lentils. Comfrey (Symphytum officinale) is best planted at the base of fruit trees, where it mines deep minerals to support tree health and serves as an excellent compost activator when chopped back. Queensland Arrowroot (Canna edulis) contributes both a rich mulch source and a delicious, starchy root crop.

As your system develops, wild herbs and opportunistic plants will naturally appear. Rather than seeing them as a problem, understand they often serve a purpose—helping to build soil fertility and heal exposed ground. However, it's important to manage these wild species before they dominate. Regular chop-and-drop before flowering will cycle nutrients while preventing them from becoming persistent weeds.

To reduce unwanted growth, aim for complete ground cover. A diverse mix of annual herbs and vegetables, along with vigorous spreading crops like Sweet Potato (Ipomoea batatas), Pumpkin (Cucurbita spp.), and Pinto peanut (Arachis Pintoi), will protect the soil, suppress weeds, and contribute to overall garden productivity.

Year 1

Year 1 - Pigeon Pea protects fruit tree from harsh sun and winds. Comfrey & Pinto Peanut support the earth with minerals and mulch. Herbs, vegetables & spices can be planted from the beginning, providing food and 'chop n drop' mulch.

Year 2

Year 2 - Pigeon Pea support the fruit tree from wind and can be chopped back for mulch. Other plants around the base still provide a yield and continue to work as a beneficial component to the system.

Year 3

Year 3 - Pigeon Pea cut back to the ground, Comfrey and Pinto Peanut still remain as support. The fruit tree will now show signs of its first fruit setting (depending on species).

Grow your own mulch & fertilizer with plants

In between each productive tree you can include the nurse (support species) trees, until the time when the fruiting tree needs the space. At this point in time, the nurse or support plants can be roughly chopped into small pieces and placed around the fruit tree to decompose, providing needed nutrients. Stacking plants in this way from ground covers and various plants ranging from small to full canopy and vines (when trees mature) mimics patterns in nature. Fungus can be encouraged by using wood chip mulch and bringing in rotten logs that contain mushrooms; mycelium is the underground network of fungus that is an essential part of any forest system.

Building a Banana Circle Mulch Pit Garden

A banana circle, also known as a mulch pit garden, is a multifunctional, highly efficient feature that belongs in every productive garden. These systems are ideal for improving soil fertility, retaining moisture, and managing large amounts of organic material - especially in sandy or heavy clay soils. They also double as a composting system and can be connected to greywater outlets for even greater benefit.

Benefits:

- Accelerates humus production through in situ composting
- Retains moisture efficiently
- Handles large or woody biomass
- Supports compost worms and soil life
- Boosts banana productivity
- Integration with greywater recycling systems

How to Construct a Banana Circle:

1. Excavate the Pit

- Dig a central pit up to 2 meters in diameter and up to 1 meter deep.
- The total circle (including mound) should span 3–5 meters across.

2. Form the Planting Mound

- Use the excavated soil to build a raised ring around the pit.
- Plant up to 7 banana plants evenly spaced around this ring.
- You can also include:
    - Taro (Colocasia esculenta) or other moisture-loving species on the inner edge
    - Sweet potato (Ipomoea batatas) or pinto peanut (Arachis pintoi) along the outer edge as a living ground cover.

3. Install Greywater (Optional)

- A 50 mm ag pipe can direct greywater from your washing machine into the pit, just below the surface
- Important: Only use plant-based, biodegradable soaps.

4. Add a Composting Core (Optional)

- Place a compost bin or mesh ring in the centre. Fill it with coarse, hard-to-compost materials like branches, woody stems, or thorny biomass.
- If the circle is small and tight with banana plants, the trunks themselves will help hold the mulch in place, making a frame unnecessary.

Why Banana Circles Work So Well

Mulch pit gardens act as a biomass reservoir, allowing you to compost materials that would otherwise be difficult to manage. The moisture-retaining pit, combined with stacked layers of organic material, creates an ideal environment for fungi, bacteria, compost worms, and beneficial insects. This accelerates decomposition—turning bulky matter into humus within months, not years.

MULCH PIT / COMPOST BIN

Planting Bananas

- Plant 100 mm deeper than the level in the pot.
- Water well and keep the soil consistently moist - but avoid overwatering.
- Leaf wetting is crucial in hot weather to prevent stress. It cools the leaves through evaporation and supports active photosynthesis. Wilted or sunburned leaves are common if this is overlooked.

Desuckering

- Remove early suckers until the mother plant reaches 2 meters in height.
- Do not keep early or shallow suckers - they're likely to topple and destabilise the main plant.
- Instead, wait about 5 months after planting to select a deep-rooted, stable sucker.
- Choose sword suckers (narrow, pointed leaves) rather than water suckers (broad, lush leaves), which produce weaker plants and smaller bunches.
- Unwanted suckers can simply be chopped and tossed into the centre of the circle to decompose.

Bagging Your Bananas

- Once the fruit fingers curve upward and the flower bell stops producing, it's time to bag.

- Clean off dead leaves or debris to prevent staining.
- Wrap the bunch in shade cloth tied at the top and bottom with twine. This protects from fruit fly, birds, and animals.
- Ensure the wrap is not too tight to avoid restricting fruit development.

Harvesting Tips

- Tree-ripened bananas taste the best. Harvest one hand at a time when they begin to yellow.
- To ripen bananas indoors, place them in a cool, dry place with a ripe banana to speed the process.

Blake kneeling next to a Dwarf Cavendish Banana bunch

Credits & further info

For nearly two decades, I have dedicated myself to the study, practice, and teaching of permaculture - both across Australia and internationally. My journey has been shaped by a deep commitment to regenerative land management, with hands-on experience in broadacre regenerative farming, biodynamics, syntropic agroforestry, market gardening, and bio-intensive, nutrient-dense food production. I have a particular passion for edible and functional perennial plants, and I find great joy in growing, harvesting, and preparing food directly from the garden.

My work is driven by a lifelong fascination with the patterns of nature and how these patterns can be applied meaningfully in our lives and landscapes. Permaculture design resonates deeply with me - it's a way of reading the land, working in harmony with its natural rhythms, and cultivating a reciprocal relationship with place. By applying regenerative principles, I create landscape designs, estimates, and DIY manuals that serve as blueprints for beautiful, edible ecosystems - spaces that nourish families, regenerate soil, and bring daily joy and connection.

Together with Tonielle Christensen of **earthmumma.co**, We co-facilitate a range of permaculture education programs, including:

- *Various earth building workshops, Including the Rocket Wood Fired Oven*
- *Introduction to Permaculture*
- *Incredible Edibles: Tropical Superfoods Workshop*
- *Permaculture Design Course (PDC)*
- *Nature Kids: Online Permaculture Program for Families*

Tonielle Christensen (left) and Blake Hudson (right)

EartheArtisan

Made in the USA
Monee, IL
05 November 2026

67383930R00029